DEMCO

Solar Electric Power Generation

Stefan C.W. Krauter

Solar Electric Power Generation - Photovoltaic Energy Systems

Modeling of Optical and Thermal Performance, Electrical Yield, Energy Balance, Effect on Reduction of Greenhouse Gas Emissions

With 107 Figures and 60 Tables

GOVERNORS STATE UNIVERSITY
UNIVERSITY PARK
IL 60466

 Springer

Author

Prof. Dr. Stefan C.W. Krauter
Rua Barata Ribeiro 370
22040-000 Rio de Janeiro RJ Brazil
email: info@stefankrauter.com

TK
2960
.K73
2006

Library of Congress Control Number: 2006921402

ISBN-10 3-540-31345-1 Springer Berlin Heidelberg New York
ISBN-13 978-3-540-31345-8 Springer Berlin Heidelberg New York

This work is subject to copyright. All rights are reserved, whether the whole or part of the material is concerned, specifically the rights of translation, reprinting, reuse of illustrations, recitation, broadcasting, reproduction on microfilm or in any other way, and storage in data banks. Duplication of this publication or parts thereof is permitted only under the provisions of the German Copyright Law of September 9, 1965, in its current version, and permission for use must always be obtained from Springer. Violations are liable for prosecution under the German Copyright Law.

Springer is a part of Springer Science+Business Media
springer.com
© Springer-Verlag Berlin Heidelberg 2006
Printed in The Netherlands

The use of general descriptive names, registered names, trademarks, etc. in this publication does not imply, even in the absence of a specific statement, that such names are exempt from the relevant protective laws and regulations and therefore free for general use.

Typesetting by the authors and SPI Publisher Services
Cover design: Estudio Calamar, Viladasens

Printed on acid-free paper   SPIN: 11587248   62/3100/SPI   5 4 3 2 1 0

for my
Parents & Norma

Preface

There is anxiety throughout the world concerning reserves of energy. The demand for electricity has increased of late years at an exponential rate, and if the demand for coal oil and gas has nearly followed a straight-line law, the slope of the line has been such as to cause concern among individual nations as to when their own supplies of fossil become exhausted, and to the world in general as to possible sources of energy when there is no more coal or oil . . . In consequence of this position, there is the greatest activity all over the world to eke out coal reserves by using other sources of energy.

From Nature 5 August 1950

1. *If the present growth trends in world population, industrialization, pollution, food production, and resource depletion continue unchanged, the limits to growth on this planet will be reached sometime within the next 100 years. The most probable result will be a sudden and uncontrollable decline in both population and industrial capacity.*

2. *It is possible to alter these growth trends and to establish a condition of ecological and economic stability that is sustainable far into the future. The state of global equilibrium could be designed so that the basic material needs of each person on earth are satisfied and each person has an equal opportunity to realize his or her individual human potential.*

3. *If the world's people decide to strive for this second outcome rather than the first, the sooner they begin working to attain it, the greater will be their chances of success.*

Résumé of "The Limits of Growth," Meadows et al. 1972

Since the article in *Nature* more than half of a century, since the résumé of the report to the "Club of Rome" more than a quarter of a century has passed. Meanwhile — after heavy discussions in the seventies and early eighties — these theses are accepted even by former vehement opponents. It even seems that these formerly so called "panic makers" are common sense nowadays. On the other hand, it looks as if we have already passed the limits mentioned – the rise of global temperature and the accumulation of natural catastrophes in the last decade are indicating it – despite better technology, stronger environmental laws, many conferences and a more ecological common sense. 1998 was by far the warmest year since worldwide records began. Mean global temperature reached a level that exceeded any of those recorded in the last 130 years and confirmed the pronounced trend (Munich Re Insurance 1998). Resignation doesn't bring us closer to the fulfillment of the needs mentioned by the report. This book has been written as a small contribution to the setting up of a sustainable development. I sincerely thank all those who helped to improve it: Dr. Paul Grunow from Q-cells, Dr. Christoph Baumann and Dr. Dieter Merkle from Springer Scientific Publications, Keith Parsons from NYU, and all the students who contributed considerably to the research projects. Particular thanks to Dr. Franz Alt and to Dr. Hermann Scheer for their encouraging forewords, and to Prof. Martin Green for his brief history on photovoltaics.

Rio de Janeiro & Fortaleza, October 2005   Stefan Krauter
                                            Prof. Dr.-Ing. habil.

Foreword by Dr. Franz Alt

TV-Journalist, editor and bestseller author.
German Solar Award 2004 and 1994, Human Rights Award 2003, Newsletter Award 2003, Environment Online Award, "Top Business Site" 2002, European Solar Award 1997, Environment Award "Golden Swallow" 1992, Siebenpfeiffer-Award 1987, Karl-Hermann-Flach Award, Drexel-Award, Hans-Thoma Medal 1983, Adoph Grimme Award 1979, Bambi TV-Award 1978.

The Solar Energy Change is Possible

At the last UN World Summit the condition of our planet has been described as follows:

- half of the world's population has to survive with less than 2 € a day
- 26,000 people are dying each day of hunger and of lack of water
- the industrialized countries consume as much coal, gas and oil per day as nature generates within 500,000 days
- each 32 hours the US is spending for military and wars as much money as the UN's annual budget
- due the greenhouse effect almost a hundred species of animals and plants are becoming extinct every day
- every day humanity grows by a quarter of a million people
- the four richest men of the US possess more money than the poorest billion of the planet

Are we insane? Can we be rescued?
This book will search for answers on that central question of the future.

We are in front of the largest moral, economic, and social challenge in human history. We are already fighting several great wars at the same time:

- wars for the last resources of our planet, e.g., oil in Iraq
- a World War against nature and thus against us

Climatic change may turn into climate collapse and thus be life-threatening for humanity. Consequences of global warming that occured just in the recent months: flooding in Bavaria, Austria, and Switzerland with billions of euros of damage, typhoons in East Asia, which caused 150 lives, the hurricanes in the south of the United States with several thousands of deaths and damages of 200 billion dollars and increasing hurricane activities in the whole Caribbean region.

The climate researchers predicted those situation amazingly accurate. We didn't want to know about it and repressed the hazard. Greed, ignorance and indolence have been stronger than comprehension, humbleness to the laws of nature and our willingness to change.

Even more dramatic for the future of the humanity and for all life on our planet is the fact that in the future all of the 6.5 billion people want to live like the 800 million in the industrial countries. We know that the planet will not bear that development. On the other hand that development cannot be stopped. The past "have-nots" will claim their fair right to live like we do.

Citizens Towards the Sun, Towards Freedom – the Solar Age Begins

The most human vision for 21st century is called the Solar Age. We are able to achieve a 100% renewable energy supply within the next 40 years. We do not need oil from the Middle East, nor gas from Siberia, nor uranium from Australia. We have all energy carriers that we shall need for the future at our doorstep: sun, wind, hydro power, geothermal energy, biofuels from farmlands and woods.

According to a study from the European Commission world's energy supply in the year 2050 could look like this:

- 40% Solar power
- 30% Biomass
- 15% Wind power
- 10% Hydro power
- 5% Oil

A path towards an economic, ecological, peaceful and forever sustainable energy supply is demonstrated in that study. Multinational oil companies such as Shell and BP have internally elaborated similar energy scenarios.

Oil, gasoline, coal and gas are getting more expensive, are destroying the environment will be depleted within a few decades. It is more intelligent to shift duly, than to further ruin the planet. It is true that the change towards solar energy will cost us some money, but no energy-change will cost us our world.

Sun, wind, hydro power, geothermal energy and biofuels are available forever for a fair price. And all will profit: The climate, the economy, the jobs, you and me, and all the more our children and grandchildren.

Unemployment rates in Germany, Europe and the World are high, the economies are stagnating – but renewable energies are a growing branch of industry. Examples:

- China installed in 2004 more than 18 million square meters of solar thermal collectors

- Japan has the world's largest photovoltaic enterprises and created hundreds of thousands jobs by them (mainly using German solar technology)

- Germany is the world leader in production and application of wind turbines. Six percent of its electricity is already generated by wind power in 2005.

- California will produce, in the reign of its governor Arnold Schwarzenegger, a third of its energy necessities via renewable energy sources and the Philippines will reach 40 percent by that time.

- Brazil is already generating more than 25 percent of its car fuel via regenerative raw materials.

Can That All be Financed?

The great economic advantage of ecological energy generation is that sun, wind, hydro power and geothermal energy never will send an invoice. The matter is available almost everywhere where we need it – without the necessity of complex global transport routes. What we need is the mass production of the new energy technologies. The prices of those have fallen by 50 percent since 1995 - and the costs of the old energies doubled by that time period. The one who is burning wood-pellets nowadays instead of using fuel oil for heating pays about half the price.

You will become acquainted more with similar surprising examples in this important book by Stefan Krauter. The message of this book and the message of the *RIO 6* conference (www.RIO6.com) is:

The Solar Energy Change is possible and the Solar Age has begun already.

Baden-Baden, October 2005   Dr. Franz Alt

www.sonnenseite.com

Foreword by Dr. Hermann Scheer

Member of German Parliament since 1980, President of the European Association for Renewable Energies, General Chairman of the World Council for Renewable Energies since 2001. Laureate of the World Solar Award 1998, Alternative Nobel Prize 1999, Word Award for Bioenergy 2000, "Hero of the Green Century" of TIME-Magazine 2002, World Award for Wind Power and Global Renewable Energy Leadership Award 2004.

Sun's True Energy Contribution

Solar energy already contributes to 94% of our planets energy use: it warms up Earth's surface and its atmosphere from space's -273.2°C to +14.5°C in average and is thus enabling all forms of life.

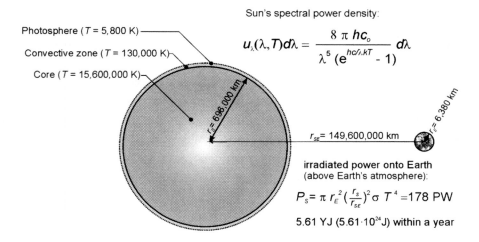

Sun's spectral power density:
$$u_\lambda(\lambda, T)d\lambda = \frac{8\pi hc_o}{\lambda^5 (e^{hc/\lambda kT} - 1)} d\lambda$$

Photosphere (T = 5,800 K)
Convective zone (T = 130,000 K)
Core (T = 15,600,000 K)

r_S = 696,000 km
r_{SE} = 149,600,000 km
r_E = 6,380 km

irradiated power onto Earth (above Earth's atmosphere):
$$P_s = \pi r_E^2 \left(\frac{r_s}{r_{SE}}\right)^2 \sigma T^4 = 178 \text{ PW}$$

5.61 YJ (5.61·10²⁴ J) within a year

Without solar energy Earth would be a dead piece of rock in space with a temperature close to absolute zero. To allow a human habitat under that conditions, we would need approximately 15 times more commercial energy than we consume today (15 · 429.4 EJ). All fossil fuel resources would be exhausted within a couple of years.

All existing fossil resources of energy such as coal, oil and gas are derived from biomass, photosynthesis and thus solar energy: Our planet has absorbed sunlight for hundreds of millions of years to create all those fossil resources that last just for 200 years of industrial human civilization.

Photovoltaics - the Most Versatile Application of Solar Energy

Solar electric power generated via the direct conversion of solar radiation into electricity – Photovoltaics (PV) – enables humanity to make use of sunlight in a clean, ever lasting, and highly versatile way. Nowadays commercial PV converts 15% of the incoming solar irradiance for at least 30 years into sustainable electricity on all parts of the planet.

This book provides an ample amount of information treating human-caused climate-change, the potential of PV to reduce greenhouse gas emissions for different scenarios (on- and off-grid, applications in the Tropics and in Central Europe) and a Life-Cycle-Analysis, including recycling of system components. The results are based on an extensive model for the calculation of the actual electricity yield of PV power plants. That model considers all optical interfaces and layers passed by the sunlight from the sun into the solar cell, thermal layout of any PV module design and its heat transfer mechanisms, its actual photo-electric conversion efficiency, allowing an accurate calculation of the yield and the optimization of PV system components, thus reducing costs for solar electricity.

The book with its tables and reference data is a valuable source of information for PV system professionals, students of physics, engineering and environment, but also for everyone interested in the subject of solar electricity.

Berlin, November 2005   Dr. Hermann Scheer

Contents

Preface VII
Foreword by Dr. Franz Alt IX
Foreword by Dr. Hermann Scheer XIII
Table of Contents XV

1   Introduction 1
    1.1   World's Energy Consumption 1
    1.2 CO_2-Emissions by Humankind 2
    1.3   Global warming by CO_2 4
    1.4   Measures of CO_2-Diminution 9
    1.5   Conventional and Renewable Sources of Energy 10
    1.7   Approach 17
          1.7.1   Production 17
          1.7.2   Yield 17
          1.7.3   Balance 17
          1.7.4   Optimization 18

2   Photovoltaics 19
    2.1   Brief History 19
    2.2   Photovoltaic Effect 21
    2.3   Photovoltaic Generator 27
          2.3.1   Electrical properties 28
                  2.3.1.1   Equivalent Electrical Circuit 30
                  2.3.1.2   Bypass Diodes 31
                  2.3.1.3   Electrical Terminals 32
                  2.3.1.4   Parallel Strings 33
          2.3.2   Mechanical Properties 34
                  2.3.2.1   Sandwich Lamination 34
                  2.3.2.2   Framing   34
                  2.3.2.3   Fixing and Mounting 35
    2.4   Properties of PV Generators in Operation
          Conditions 37
    2.5   Mounting of PV Modules 38
    2.6   Future Development of Photovoltaics 41
    2.7   Research Funding for Photovoltaics 43

	2.8		Market Development of Photovoltaics 44
3	Inverters 49		
	3.1	Autonomous Operation 49	
	3.2	Inverters for Electrical Grid Injection 50	
	3.3	Types of Inverters 54	
		3.3.1	External Commutated Inverters 54
		3.3.2	Self Commutated Inverters 54
		3.3.3	Inverters Based on PWM 55
	3.4	Electrical Grid Connection 56	
		3.4.1	Voltage Levels of Electrical Grids 56
		3.4.2	Boundary Values of Electrical Grids 56
		3.4.3	Long-Distance Transport of Electricity 57
4	Storage 61		
	4.1	Lead Sulphide Acid Battery. 62	
		4.1.1	Principle 62
		4.1.2	Gassing 63
		4.1.3	Specific Gravity (SG) 63
		4.1.4	Operating Temperature 64
		4.1.5	Self-Discharge 65
		4.1.6	Deep Discharge 66
		4.1.7	Sulfation 67
		4.1.8	Battery Types 67
	4.2	Other Type of Batteries 70	
		4.2.1	Nickel Cadmium Battery 70
		4.2.2	Nickel Hydride Batteries 71
		4.2.3	Lithium-Ion Batteries 71
	4.3	Fuel Cells 74	
		4.3.1	Principle 74
		4.3.2	Types of Fuel Cells 74
5	PV-Systems in the Tropics 77		
	5.1	Pre-installation Issues 77	
		5.1.1	Additional Considerations for Planning 77
			5.1.1.1 Determination of Load Requirements..................... 77

			5.1.1.2 Dynamics of Project Development vs. Time Constrains. 78

		5.1.2	Financing 78
		5.1.3	Importation 78
		5.1.4	Language Barriers 79
	5.2	Technical Issues 80	
		5.2.1	Mounting 80
			5.2.1.1 Fixation of PV Modules 80
			5.2.1.2 Wiring of PV Generator 80
			5.2.1.3 Theft Prevention 81
			5.2.1.4 Safety Considerations 81
		5.2.2	Non-MPP Operation of PV Generator. 81
		5.2.3	Energy Storage 82
			5.2.3.1 Battery Types 82
			5.2.3.2 Nominal Voltage Level. 82
		5.2.4	Power Conditioning Equipment. 83
			5.2.4.1 Switching Devices 83
			5.2.4.2 Ventilation 83
			5.2.4.3 Charge-Controllers 83
	5.3	Operation and Maintenance 84	
		5.3.1	Pollution & Degradation of System Components. 84
		5.3.2	Monitoring 84
		5.3.3	Further Recommendations 85
	5.4	Concluding Remarks for PV in the Tropics. 86	
6	Energy Consumption for the Set-up of a PV Power Plant. 87		
	6.1	Preliminary Remarks 87	
		6.1.1	Differentiation of the Model Cases 87
		6.1.2	Equivalent Primary Energy Consumption 89
	6.2	Preparation of Raw Materials for Production 89	
		6.2.1	Development of a Deposit 89
		6.2.2	Release (Exploitation) 90
		6.2.3	Transport 91
		6.2.4	Preparation for Production 92
		6.2.5	Construction Work and Buildings 93
	6.3	Direct Energy Consumption at the Production Process 94	

6.4	Production of Solar Cells 95	
	6.4.1 Production of Technical Silicon (MG-Si) 95	
	6.4.2 Metallurgical-Grade Silicon (MG-Si) to Semiconductor-Grade Poly-Silicon (EG-Si)...95	
	6.4.3 Production of Single-Crystalline Silicon 96	
	6.4.4 Semiconductor-Grade Silicon to Multi-Crystalline Silicon 97	
	6.4.5 Production of Silicon Wafers (single- and multi-crystalline) 98	
	6.4.6 Single-Crystal Wafers to Single-Crystalline Solar Cells 100	
	6.4.7 Multi-Crystalline Wafers to Multi-Crystalline Solar Cells 101	
	6.4.8 Production of Amorphous Silicon Solar Cells... 102	
	6.4.9 Production of Solar Cells Made of Other Semiconductors 102	
6.5	Production of PV Modules 105	
	6.5.1 Lamination Process 106	
	6.5.1.1 Integrated Laminator 106	
	6.5.1.2 "Passing-Through" Laminator 108	
	6.5.2 Production of "Encapsulated" PV modules... 110	
	6.5.3 Production of "Laminated" PV modules 110	
	6.5.4 Electrical Power Conditioning.............. 111	
	6.5.5 Support Structure 112	
6.6	Installation and Taking into Operation 113	
	6.6.1 Transport 114	
	6.6.2 Installation 114	
	6.6.3 Setting into Operation 114	
6.7	Operation Expenses 115	
	6.7.1 Cleaning 115	
	6.7.2 Maintenance 115	
	6.7.3 Use of Land 116	
6.8	Dismantling 116	
	6.8.1 Dismantling 116	
	6.8.2 Transport 116	

7	Energy Yield		117
	7.1	Model to Determine the Cell Reaching Irradiance	117
		7.1.1 Sun's Position Relative to Earth's Surface	117
		7.1.2 Way of Sun's Irradiance Through the Earth's Atmosphere	121
		7.1.2.1 Solar Constant	121
		7.1.2.3 Direct Irradiance	124
		7.1.2.4 Diffuse Irradiance	124
		7.1.2.5 Albedo	128
		7.1.2.6 Angular Distribution of Yearly Irradiance in Central Europe	128
		7.1.3 Optical Model of Module Encapsulation	129
		7.1.3.1 Optical Interface at Boundary Layers	131
		7.1.3.2 Optical Transmittance of a Plane Slab	134
		7.1.3.3 Internal Transmission and Reflection	136
		7.1.3.4 Transmittance Through Two Slabs	137
		7.1.3.5 Transmittance Through Three Slabs	140
		7.1.3.6 Optical Transmittance Through m Slabs	141
		7.1.3.7 Simulation Results	142
	7.2.1	Heat Flow Input	147
		7.2.1.1 Heat Flow Input by Sky and Ground Radiation	147
		7.2.1.2 Heat Flow Input by Ambient Temperature	148
		7.2.1.3 Heat Flow Input by Irradiance	148
	7.2.2	Heat Transfer Inside a Module	150
		7.2.2.1 Dimensional Layout of the Thermal Model	150
		7.2.2.2 Stationary Heat Flow in the Module	151
		7.2.2.3 Non-Steady-State Heat Flow in the Module	152
	7.2.3	Heat Dissipation	153
		7.2.3.2 Determination of Sky Temperature	156
		7.2.3.3 Heat Dissipation by Natural Convection	157
		7.2.3.4 Heat Dissipation by Forced Convection	160

			7.2.3.5 Heat Transfer for Superposition of Natural and Forced Convection........... 161
		7.2.4	Model Calculation 163
		7.2.5	Validation of Thermal Modeling 164
	7.3	Electrical Modeling 166	
		7.3.1	Current 166
		7.3.2	Other Electrical Parameters 167
	7.4	PV Grid Injection 168	
		7.4.1	Modeling of Inverters 168
		7.4.2	Limiting Factors for the Design of PV Power Plants 169
	7.5	System Layouts 170	
	7.6	Electrical Yield of a Reference System 172	

8	Energy Input by Dumping and Recycling 173	
	8.1	Separation of Materials 173
	8.2	Energy Input by Recycling 174

9	Total Energy Balance 177		
	9.1	Commutated Energy Expense 177	
	9.2	Models for Energy Balances 178	
	9.3	Input-Output Analysis 179	
	9.4	Process Chain Analysis 180	
	9.5	CO_2 Reducing Effects by the Use of PV 182	
		9.5.1	Specific Emission Balance 183
		9.5.2	Effect of PV on Reduction of CO_2 Emissions in Germany........................ 184
		9.5.3	Variation of Location 186

10	Optimization 191		
	10.1	Improvement of Irradiance on a Solar Cell.......... 192	
		10.1.1	Improvement of Irradiance by Tracking the Sun 192
		10.1.2	Improvement of Cell Irradiance by Reduction of Optical Reflection.................. 192
			10.1.2.1 Structuring of the PV Module Surface 193
			10.1.2.2 Selective Structuring 193
			10.1.2.3 Improved Matching of the Refractive

			Indices of the Module Encapsulation Layers 196
		10.1.2.4	Additional Anti-Reflective Coating.... 197
10.2	Reduction of Expenses for Mounting 198
10.3	Substitution of Building Components 199
	10.3.1	Solar Roof Tiles 199
	10.3.2	Solar Facades 201
10.4	Thermal Enhancement of PV Modules 203
	10.4.1	Real Operating Cell Temperatures Under Tropical Conditions 203
	10.4.2	Preliminary Work for the Reduction of Temperatures in PV Modules 205
	10.4.3	Development of a Thermally Improved Prototype 205
	10.4.5	Construction, Operation and Measurement of TEPVIS in Africa 208
	10.4.6	The Integrated Solar Home System (I-SHS)... 211
		10.4.6.1 Composition of the System 211
		10.4.6.3 Benefits of the I-SHS 215
		10.4.6.4 Further Development 216

11 Summary 217

12 Appendix 221
    12.1 List of Symbols and Abbreviations 221
    12.2 Tables 229

Literature 259

1
Introduction

1.1 World's Energy Consumption

By using fossil fuels such as wood, coal, oil and gas, it was possible for humanity to set up civilization in colder climates. Due to the increasing demands of comfort, a higher mobility and a larger world population, energy consumption rose tremendously over the last 150 years (see Figure 1.1) and exhaustion of these fuels is foreseeable in the midterm.

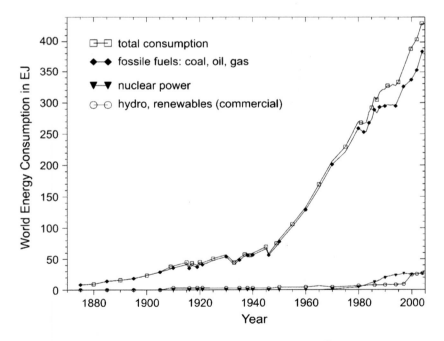

Fig. 1.1. Primary energy consumption of the world in EJ (10^{18} Joule) as a function of time (1875-1995) with shares of CO_2-emitting and CO_2-free fuels (Sources: 1875-1965: Interatom-Shell-Study 1992, 1970-1988: UN Yearbook of World Energy Statistics, 1989-2004: BP Statistic Review of World Energy, 1996)

Additionally, the emitted carbon dioxide hinders the heat radiation exchange between the Earth's surface and space, which causes climatic changing effects (see following chapters). While these facts have been known since the early seventies (Meadows et al. 1972), energy consumption of humanity (and related CO_2-emissions) rose to 429.4 EJ ($429.4 \cdot 10^{18}$ J) in the year 2004.

Figure 1.2 illustrates the potential of solar energy: the Sun's irradiation on Earth is 14,000 times higher than the World's energy consumption. Accumulated over one year, the energy of solar irradiance on Earth is much higher than all known fossil fuel resources.

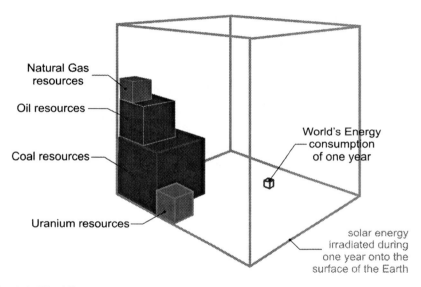

Fig. 1.2. World's energy consumption in comparison to all its fossil resources and its annual solar energy potential (adapted from Greenpeace).

1.2 CO_2-Emissions by Humankind

Human energy needs have been fulfilled by burning such fossil fuels as coal, oil and gas, which accordingly have led to elevated CO_2-emissions, especially since the beginning of industrialization as shown in Figure 1.3. Conversion back from CO_2 to O_2 by photosynthesis cannot be done entirely by the amount of plants (biomass) presently existing; thus an accumulation of CO_2 in the atmosphere is observed (see Figure 1.4). This effect is enhanced by a reduction of the amount of plants (e.g., due to deforestation).

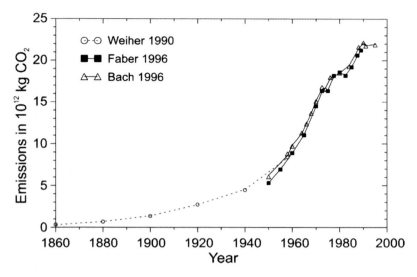

Fig. 1.3. Global anthropogenous CO_2-emissions as a function of time (1860-1995) according to Weiher 1990, Faber 1996, Bach 1996, Worldwatch 2000.

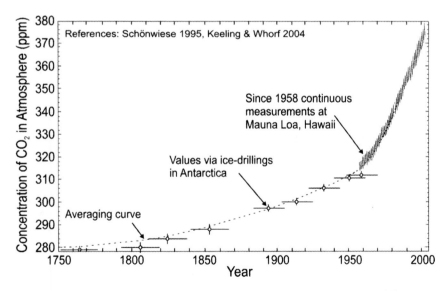

Fig. 1.4. CO_2-contents of the Earth's atmosphere in ppm ("parts per million") as a function of time. Before 1958 detection by drillings in the Antarctica: range of confidence indicated; from 1958 onwards continuous measurements at Mauna Loa (Hawaii). Graphics based on Schönwiese 1995, Keeling & Whorf 2004.

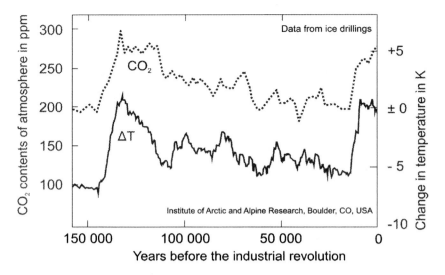

Fig. 1.5. Correlation of CO_2-contents of the Earth's atmosphere and its temperature change from 150,000 B.C to 1750 A.D. according to the Institute of Alpine and Alpine Research, Bolder, Colorado, USA, published in National Geographics 2000

1.3 Global Warming by CO_2

While the main components of the atmosphere (N_2 and O_2) allow the same good optical transmittance of incoming solar irradiation as for the heat radiation from the Earth's surface to space, the gases relevant for the greenhouse effect (such as water-vapor, methane, N_2O, and ozone) show a good transmittance just for the visible part of the radiation (λ = 350–800 nm), but hinder the emission of infrared heat radiation (λ > 10,000 nm) from earth to space. An equilibrium of incoming and outgoing energy flows then occurs when the earth surface is radiating more, but this occurs at a higher surface temperature - the greenhouse effect. Without this natural greenhouse effect, the surface of the earth would be about 30 K colder. An overview of the possible effects of natural greenhouse gases is given in Table 1.1.

Table 1.1. Components of natural greenhouse effect (Schönwiese 1995)

Gas, chemical formula	Share of natural increase of temperature	Relative share
Water vapor, H_2O	20.6 K	62%
Carbon dioxide, CO_2	7.2 K	22%
Ozone near ground, O_3	2.4 K	7%
Nitrous oxide, N_2O	1.4 K	4%
Methane, CH_4	0.8 K	3%
Others	ca. 0.6 K	2%
Sums of shares	33 K [1]	100%

[1] Alternative estimations are showing a total effect of 15–20 K only; investigations of the Intergovernmental Panel on Climate Change (IPCC 1994) are showing an effect of 30 K (incl. clouds).

Human activity has caused an increase in the emissions of natural and synthetic greenhouse gases, notably since the beginning of the industrial age. As a result, greater amounts of infrared heat radiation are trapped in the atmosphere which then causes an increase of ground surface temperature. The atmosphere's reflection of infrared radiation emitted from earth has increased by about 1% since 1850 (or by 3 W/m² over the natural back-reflection of 320 W/m², see Fischedick et al. 1999).

Climate change as a result of an elevated contents of carbon dioxide in the atmosphere had already been postulated in 1896 by the Swedish chemist Svante Arrhenius (see Arrhenius 1896). Astonishingly he predicted the dimension of the greenhouse effect by 5 K at a doubling of the CO_2-contents quite accurately. Interest in climatic research was aroused in 1938 when the British chemist Callendar showed an increase of the atmospheric carbon dioxide during the past decades. Nevertheless, an international focus on carbon dioxide only started in 1971–72 when it was recognized, that its effects could be as severe as general air pollution which had been at the foreground of discussion at the time.

In the year 1977, the World Meteorological Organization, an UN-Organization located in Geneva, Switzerland, called a commission of experts, who declared the need for a world climate congress. In 1970 that congress took place and triggered international research which resulted in a vast increase of knowledge about the climate – most specifically the mechanisms of climatic change and the greenhouse effect.

Recent results of emissions effect on the climate are given in Table 1.2. Evidently anthropogenous CO_2 with a total effectivity of 61% (50% according to Flohn 1989) has the largest effect on global warming, while methane at 15%, FCCHs at 11% (20% according to Flohn 1989) and ozone at 9% (10% according to Flohn 1989) contributes to a smaller extent.

Table 1.2. Concentration of greenhouse gases in the atmosphere and climatic efficiency of anthropogenous emitted trace gases (Schönwiese 1995, IPCC 2001)

Gas, chemical formula	Concentr. in ca. 1800 (in ppm)	Concentr. in 1991 and 2005 (in ppm)	Human-caused emissions (in Mt/a)	Avg. remain time (in a)	Relative molar green-house effectivity [1]	Share on total effect [3] (in %)
Carbon dioxide, CO_2	280	355 (1991) 380 (2005)	29,000	5 to 10 [4]	1	61
Methane, CH_4	0.8	1.7	400	10	11 (23) [2]	15
CFC, CFM, Freon	0	0.00025 to 0.00045	1	55 to 115	3,400 to 7,100 (12,000) [2]	11
Dinitrous oxide, N_2O	0.29	0.31	ca. 10	130	270 (296) [2]	4
Ozone at ground, O_3	unknown	0.015 to 0.05	ca. 500	0.1 to 0.25	unknown	9 [5]

[1] Relative molar greenhouse efficiency assuming a 100-years time horizon.
[2] Values in brackets represent the latest results as published by IPCC 2001.
[3] Contribution to anthropogenous greenhouse effect for a 100-y time horizon.
[4] Anthropogenous effective time 50 to 200 years.
[5] incl. all other relevant trace gases.

The climatic efficiency of the greenhouse effect can be validated by measurements of air temperature. Despite fluctuations, a distinct increase of the global air temperature at sea level by 0.5–1.0 K has occurred since the end of last century, as seen in Figure 1.6. Besides meteorological effects (winds and currents of the oceans, sea- and groundwater-level), this increase also causes changes of biological activity [1] that may result in major changes in the human habitat.

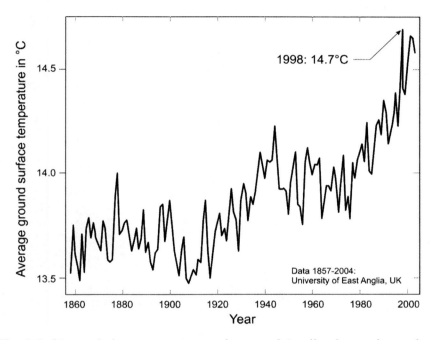

Fig. 1.6. Observed air temperature near the ground (mediated over the northern hemisphere) from 1858 to 2004.

It has also to be considered that the actual increase of temperature turns out to be lower than it should be according to the increase of atmospheric CO_2. This is due to a temporary temperature decrease caused by anthropogenous tropospheric sulfide (Newinger 1985, Charlson et al. 1992, Kiehl et al. 1993, Kaufmann et al. 1993, Charlson et al. 1994). The

---

[1] an increase of 10 K in temperature results in a doubling of the speed of chemical-biological reactions (RGT-rule), until an upper limiting temperature (ca. 60°C for enzymes), see Linder 1948/1977.

consequences of the increase in atmospheric temperature are not only limited to an increased frequency of natural disasters such as floods and hurricanes, see Figure 1.7, but also enhanced chemical and biological activities such as corrosion of buildings, faster growth of bacteria and the spread of disease transmitting animals. To limit such occurrences, man-made CO_2-emissions should be reduced either by a reduction in energy consumption or by adopting energies that emit less CO_2.

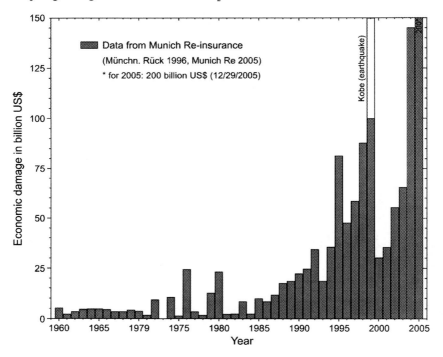

Fig. 1.7. Worldwide damage (in billions US$) caused by natural disasters as a function of time (Munich Re 2005, Berz 2003, Münchn. Rück 1996).

Yet the best solution is a combination of both strategies: energy conservation and substitution of conventional energy resources (see Fig. 1.8). Without the adoption of such policies, currently respected forecasts of anthropogenic climate change predict an increase in global mean temperatures above the pre-industrial era by 1–2.5 K by the decade of 2036 to 2046. Note, this range is relatively prone to errors in the model's climate sensibility, rates of oceanic heat uptake or global response to sulphate aerosols as long as these errors are persistent over time (Allen et al. 2000).

1.4 Measures of CO_2-Diminution

Facing the consequences of an accumulation of CO_2 in the atmosphere, the German government – as well as many other governments – decided to reduce the CO_2 emissions by 50% by the year 2020. This does not seem achievable by the means being used (i.e., merely by an increase of efficiency of power generation). While the specific crude energy demand fulfilling a certain gross national product was successfully reduced, this action alone will not meet the goal of a 50% reduction to be reached by 2020. Only if a considerable amount of energy can be generated with less emission of CO_2 can this aim be fulfilled. In a long-term perspective this means a substitution of fossil fuel power plants by renewable energy converters. Figure 1.8 shows that a reduction of CO_2 emissions becomes more effective through a combination of economic measures and replacement of CO_2 emitting power plants. According to the objectives mentioned above, future power plants have to be renewable [2]. Renewable energies are emission-free, are almost of unlimited availability, bear negligible secondary costs, their cost-trend is digressive and they have a good social acceptance by the population. In this book the use of photovoltaics for electrical energy generation is discussed; as an important example for renewable energy conversion.

---

[2] The use of nuclear power will not be considered as an alternative as mentioned in the treatment of the following chapter.

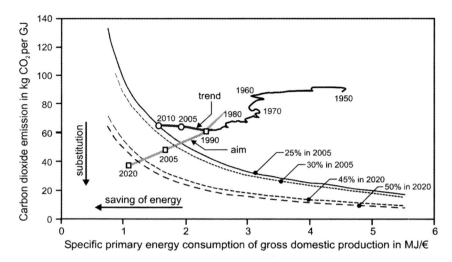

Fig. 1.8. Reduction of CO_2-emissions by depletion of energy consumption and by substitution of CO_2-intense energy generation. Also displayed are goals for CO_2 reductions.

1.5 Conventional and Renewable Sources of Energy

In the following section, the carbon based fossil sources of energy (oil, gas and coal) and nuclear power will be classed as "conventional sources of energy."

Fossil fuels are a product of the photo-synthetic process, which occurred many millions of years ago. Simply put they can be considered as none other than stores of solar energy or solar radiation. It has taken more than 100 million years to obtain the existing forms of fossil fuels and thus their formation is to be considered as a geological one-time event with extremely low conversion efficiency, see Table 1.3. From a human perspective, fossil resources must be looked upon as limited. The concept "renewable" hardly applies to them.

Table 1.3. Energy conversion time scale and conversion efficiencies of solar energy into different energy carriers

Energy carrier	Time for "production" of energy in years	Solar conversion efficiency	Literature
Coal, lignite	> 150,000,000	< 0.001%	Bennewitz 1991
Oil, gas	> 100,000,000	< 0.001%	Bennewitz 1991
Wood	1–30	1% [3] 0.55% 0.1% [4]	Kaltschmitt 2003 Kleemann 1993 Spreng 1995
Biomass photosynthesis	0.1–1	0.3% – 5% [5] 0.04% – 1.5% 0.2% [6]	Kleemann 1993 Kaltschmitt 2003 Spreng 1995
Hydro power	0.01–1	< 1%	
Wind power	continuously	0.25% 2%	Hoagland 1996 Kleemann 1993
Photovoltaics	continuously	6% – 25%	Green 1995

The present price advantage of fossil fuels justifies itself for favorable political considerations as the price war of the oil exporting countries (OPEC), their "terms of trade" toward major countries of consumption and the direct and indirect subsidization of conventional fuels. Direct subsidization is done by giving subsidies for exploration, mining and transport: e.g., for coal and nuclear power in Germany and Diesel in Brazil. Indirect subsidization is done by charging the population and the government for the follow-up costs (e.g., air and water pollution control, security of supply for oil, security for nuclear waste). For example, measures to secure oil supply, such as military presence (e.g., in Saudi Arabia) or even

---

[3] i.e. beech wood: irradiance 3.7 PJ/(km^2 a), storage in dry mass above ground: 570,000 kg (240,000 kg below the ground as roots and humus).
[4] from avg. irradiance and energy density of forest growth for moderate climates.
[5] maximum of solar yield is 5.4% for sugar beets (farmland in general: 0.3%).
[6] photosynthesis related to the global average.

direct intervention (e.g., Iraque – the last war there did cost the US tax payers about 300 billion US$). Another example: Plutonium loses only 50% of its activity in 12,500 years; a 24-hrs observation by just one guard will cost 900 million US$ during that time. The price trend of conventional fuels in the medium-term fluctuated greatly (Fig. 1.9), but as a restricted commodity with constant (or even increasing) demand, its price may increase in the long term.

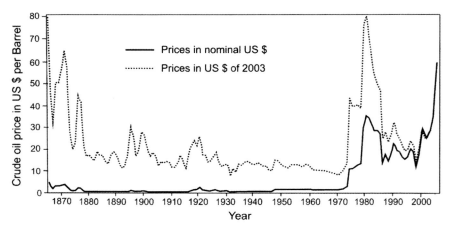

Fig. 1.9. Development in price of crude oil in US $ (actual and 2003 value) during the last 150 years.

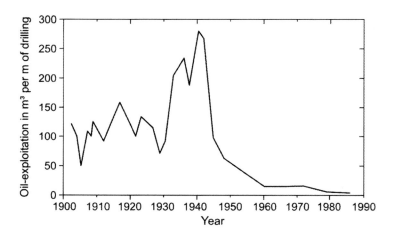

Fig. 1.10. Exploitation rate of US oil-well drillings since 1900.

An interesting aspect can be observed in the price development: in the same time period when the exploitation rate of oil drillings in the US decreased (to virtually zero) in the mid seventies (see Fig. 1.10), the oil price on the world market exploded (see Fig. 1.9), a similar (but less severe) effect could be observed in the mid forties.

Social effects caused by using these sources of energy must be considered also, such as increasing health impairments (respiratory tract illnesses, allergies etc.) and destruction of cultural possessions and environment (acid rain), too (see Hohmeyer 1989). Destruction caused by air pollutants is exemplified by the damage to historical buildings and monuments in Munich (shown in Figure 1.11): From 1700 to 1850, the time taken to increase by one grade of damage by air pollutants was calculated to be more than 300 years; this dropped suddenly to less than 50 years, in the period between 1930 and 1955. Today the time to increase one grade of damage is between 70 and 120 years, depending on the time of the building's construction. This means a doubling for the costs of restoration compared to 150 years ago, although the pollutants already have been reduced to some extent. The accumulated costs for additional restorations of cultural monuments in Germany are approx. US$ 70 billion for the year 2000. For buildings without cultural value (structures, bridges, industrial plants, high voltage transmission towers) the costs for additional maintenance due to damage by air pollutants are about US $ 4.1 billion every year in Germany, according to a study of the Federal Institute of Material Research (BAM 1990).

The use of nuclear power results in non-reversible technical and administrative (also political and social) structures, which are contradicting democratic culture [7]. A study by the University of Münster (see Ewers et al. 1991) reveals costs of US $ 2.35 trillion for a severe accident in Germany by a "Biblis" type reactor.

---

[7] Highly radioactive nuclear waste has half-life periods of several thousand years (e.g. Plutonium: 12,500 years) and must be reliably guarded and supervised over many such half-lifes. The consequent technical, administrative and military structure is irreversible even by public demand. Thus this complex will be immune to democratic rules.

14   Solar Electric Power Generation

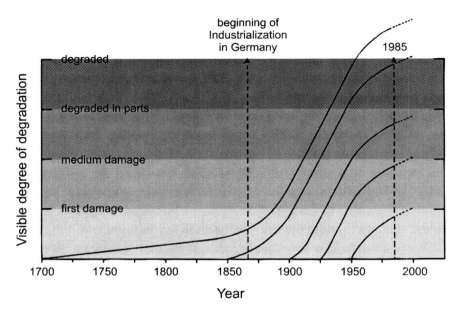

Fig. 1.11. Damage of historical buildings in Munich (Germany) as a function of time according to the *Preservation Office for Historical Monuments of the State Government of Bavaria*. Graph by Grimm et al. 1985.

The actual insurance for such an event is limited to US $ 294 million, so the costs for the insurance are underestimated by a factor of at least eight thousand. Since this form of energy use is in principle different from all other forms, and is not also momentarily ascertainable and comparable, the use of nuclear power will not be further considered at this account.

1.6 Energy Conversion

Different examinations have been already published on the energy requirement of power station facilities and the funds required for the operation with fossil and regenerative sources of energy (Aulich 1986, Schäfer 1988, Jensch 1988, Hagedorn 1989, Real 1991, Cap 1992, Spreng 1995, etc.). A short overview of the results is given in Table 1.4, see also Table A3.

Table 1.4. Emission of carbon dioxide for the conversion of fossil and renewable energy sources into electricity (in g of CO_2 per generated kWh of electricity) according to literature from 1991 to 2005.

Reference Fuel	Faninger 1991[1]	Stelzer et al. 1994[2]	DB 1995	DB 1995[3] (CO_2-equiv.)	Voß 1997	Kalt-schmitt 2003 (CO_2-equiv)	Alsema et al. 2005[4]
Wood	940						20
Coke	960						
Briquette	910						
Lignite	890		1,135.6	1,146.6			
Coal	860	830–840	917.7	1,049.0	878.4–881.3	839	1,000
Fuel oil (light)	720						
Natural gas	480					399	400
Wind power		8.0–16.3	0.7	1.1	8.1–35.7	23–48	8
Hydro power		100	1.4	1.8		10–21	
Photovoltaics (PV)		230–300	51.7	61.2	206–318	123–279	26–41

[1] Details of Faninger 1991 are based on details concerning values for light fuel oil or natural gas of Cap 1992.
[2] Details from Stelzer et al. 1994 are partial based on studies in 1993.
[3] The survey commission of the German Bundestag (ref. DB 1995) also considered the CO_2 equivalent of other climatic relevant gases. For PV, CdTe technology was considered, according to GEMIS 1992.
[4] Life-cycle greenhouse gas emissions, for wood LCA of biomass is considered, for photovoltaics grid-connected PV at 1,700 kWh/(m²a) irradiation.

It is clearly visible that during the last decade the relative emissions for solar electricity via photovoltaics sank from 230–318 g/kWh to 26–41 g/kWh due to considerable improvements in the production technology and use of materials. The very latest development indicates that, triggered by a shortage in the silicon supply, a reduction of standard wafer thicknesses from 0.3 mm to 0.2 mm will take place, therefore the values of latest publication by Alsema et al. 2005 will probably be underrun by 25–30%.

While the absolute values of the specific energy requirements and greenhouse gas emissions of PV technology may change during the years, this book should still keep its value, because the accounting method presented will still be valid in the future. Further information is given in the later chapters and in the Annex.

In the past, improprieties occurred with respect to energy amortization times. Often, the required operating fuels, such as combustibles, have not been included in the considerations. A simple coal power station, for example, achieved an "energy amortization time" of one year, in comparison to a PV power station with four years. However, utilizing the upper definition, a primitive campfire on the ground would have the best amortization time of all power stations. It is in this way that renewable energy sources have been discredited (either by purpose or by ignorance) for some time. Any energy conversion facility operated by combustible fuels has an infinite energy amortization time! Every renewable energy conversion facility works with renewable fuels, so it does not need to be counted.

Concerning the greenhouse effect, the main issue is not only energy consumption, but the effective emission of CO_2 during a complete life-cycle of a power station, its components and materials, including recycling. For example, although aluminum production has an energy consumption that is ten-times higher than stainless steel; nevertheless this may be acceptable, if this energy is generated by renewable energy sources (as it is for aluminum production in the Scandinavian countries) and the aluminum recycled later on, which allows recovery of 90% of the energy used, thus diminishing effective CO_2 emissions. (E.g., in Brazil more than 90% of the aluminum cans are recycled, see also Tables 8.1, & 8.2).

1.7 Approach

The aim of this account is to examine, how the massive use of PV generators affects the net CO_2 emission of the population. To achieve that, the complete life cycle of the PV-generator has to be considered, including factors such as production, transport, mounting, use, electrical yield and dismantling efforts.

1.7.1 Production

Besides observing the present state-of-the-art production methods, other measures that are leading to more environmentally sound production of PV-systems are examined. Here particularly the diminution of the expenditure of energy (at the same yields) and the CO_2 emissions are decisive.

1.7.2 Yield

The electrical energy generated by a photovoltaic power plant will be examined, considering all relevant parameters such as the location (irradiance, reflective losses, micro climates) and the possible interactions of these parameters.

1.7.3 Balance

The specific reduction of CO_2 through utilizing a PV system will be examined by Life Cycle Analysis (LCA). Besides being an adequate method of production (minimization of cycles of energy and matter), the system has also to increase the yields without considerable effort, if possible. New in this book is the approach by an integral analysis of a complete system, considering the origin of its components under inclusion of the recycling ability, and the operating conditions.

1.7.4 Optimization

The objective is the development of improved PV systems with consideration of the real environment and impacts on operation (irradiation, reflection, outside temperature, wind speed), and the interaction parameter of the single components with a view towards the optimization of the energetically weighted effectiveness. By setting up a prototype, the statements made will be checked. The possibilities of mass production and the resulting effect on the CO_2 balance will be examined.

2
Photovoltaics

Photovoltaics (PV) is the direct conversion of radiation into electricity. While electricity is to be increasingly used as a source of energy, photovoltaics will play an important role in the field of renewable energies. PV technology is modular (i.e., existing systems are expandable), has a long lifetime (manufactures give guarantees of up to 25 years), is silent and emissions-free during use. There is a considerable potential for cost reduction due to known semiconductor technology; in addition, present production processes can be made more efficient and cheaper developing mass production techniques. Although at present (2005), prices at the factories are stabilized in the vicinity of 3 €/W_p, due to a lack of silicon as a raw material. It is expected after new production facilities for silicon are into operation in 2006 prices will fall again. Latest information indicates that production costs of 1–1.6 €/W_p are feasible.

2.1   Brief History (Green 2000)

Solar cells derive their origin from some of the most important scientific developments of the 20th century, combining the Nobel prize winning work of several of the most important scientists of that century. The German scientist, Max Planck, began the century engrossed in the problem of trying to explain the nature of light emitted by hot bodies, such as the sun. He had to make assumptions about energy being restricted to discrete levels to match theory and observations. This stimulated Albert Einstein, in his "miraculous year" of 1905 to postulate that light was made of small "particles," later called photons, each with a tiny amount of energy that depends on the photon's color. Blue photons have about twice the energy of red photons.

Infrared photons, invisible to the eye have even less energy. Ultraviolet photons, the cause of sunburn and skin cancer, are also invisible but carry even more energy than the blue ones, accounting for the damage they can do. Einstein's radical suggestion led to the formulation and development of quantum mechanics, culminating in 1926 in Erwin Schrödinger's wave equation. Wilson solved this equation for material in solid form in 1930. This allowed him to explain the difference between metals, good conductors of electricity and insulators; also the properties of semiconductors with their intermediate electrical properties. Electrons, the carriers of electrical charge, are free to move around in metals, allowing electrical currents to flow readily. In insulators, electrons are locked into the bonds holding the atoms of the insulator together. They need a jolt of energy to free them from these bonds, so they can become mobile. The same applies to semiconductors, except a smaller jolt is needed – even the red photons in sunlight have enough energy to free an electron in the archetypical semiconductor, silicon. Russel Ohl discovered the first silicon solar cell by accident in 1940. He was surprised to measure a large electrical voltage from what he thought was a pure rod of silicon when he shone a flashlight on it. Closer investigation showed that small concentrations of impurities were giving portions of the silicon properties dubbed "negative" (n-type). These properties are now known to be due to a surplus of mobile electrons with their negative charge. Other regions had "positive" (p-type) properties, now known to be due to a deficiency of electrons, causing an effect similar to a surplus of positive charge (something close to a physical demonstration of the mathematical adage that two negatives make a positive).

William Shockley worked out the theory of the devices formed from junctions between "positive" and "negative" regions (p-n junctions) in 1949 and soon used this theory to design the first practical transistors. The semiconductor revolution of the 1950s followed, which also resulted in the first efficient solar cells in 1954. This caused enormous excitement and attracted front-page headlines at the time.

The first commercial use of the new solar cells was on spacecraft, beginning in 1958. This was the major commercial application until the early 1970s, when oil embargoes of that period stimulated a re-examination of the cells' potential closer to home. From small beginnings, a terrestrial solar cell industry took root at this time and has grown rapidly, particularly over recent years. Increasing international resolve to reduce carbon dioxide emissions as a first step to reigning in the "Greenhouse Effect," combined with decreasing cell costs, sees the industry poised to make increasing impact over the first two decades of the new millennium.

2.2 Photovoltaic Effect

A solar cell is a large-area semiconductor diode. It consists of a *p-n* junction created by an impurity addition (doping) into the semiconductor crystal (consisting of four covalent bonds to the neighboring atoms for the most commonly used silicon solar cells). If impurities are phosphorus-atoms, which have five outer electrons, only four electrons are required to fit the atom into the silicon crystal structure, the fifth electron is mobile and free. So in this region of the crystal there are many (a majority) free negative charges, therefore it is called *n*-region. Vice versa for the *p*-region: By doping the crystal with boron atoms, which have only three outer electrons, one electron is always missing for a complete binding into the crystal structure. This electron could be "borrowed" from neighboring atoms, so the place of the missing electron is shifted. This missing electron could also be seen as a "hole" with a positive charge that is mobile and wandering. There are much more free holes than free electrons in the *p*-regions, so the electrons are called minority charge carriers there.

Due to the differences in concentration at the "frontier" between the two regions, electrons diffuse into the *p*-regions and "holes" into *n*-regions, therefore an electrical field in the formerly electrical neutral junction comes into existence (see Figure 2.1): The buildup of the space-charge-region. It increases until a further practical diffusion of carriers is avoided by it.

Light (or sun radiation) falling into the semiconductor generates electron-hole-pairs, causing an increase in the concentration of the minority charge carriers by several orders of magnitude. These charge carriers diffuse to the space charge zone and are divided by the electric field there. Between the contacts of the *n*-side and *p*-side a tension V could be detected, as shown in Figure 2.1. When a load resistor R is applied, a current I flows through it, and electrical power is dissipated.

22 Solar Electric Power Generation

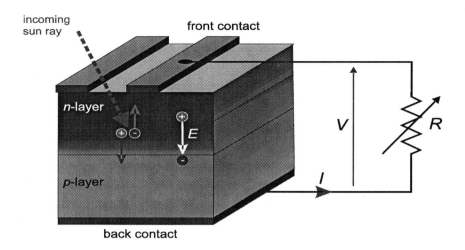

Fig. 2.1. Principle of photovoltaic energy conversion in a n-p-doped semiconductor. Generated power is supplied to an ohmic load R (scheme).

Fig. 2.2. Front view of a square multi-crystalline silicon (mc-Si) solar cell at a size of 10 cm x 10 cm.

The characteristic of a solar cell without any irradiance (dark characteristic) corresponds to a diode characteristic [8], as shown in Figure 2.3 below. When the solar cell is illuminated, this characteristic shifts by the amount of the photo current I_{phot} in blocking direction (light characteristic). This solar cell characteristic is determined by connecting a variable load resistor to it (see Figure 2.1) and plotting the resulting currents and voltages at different loads.

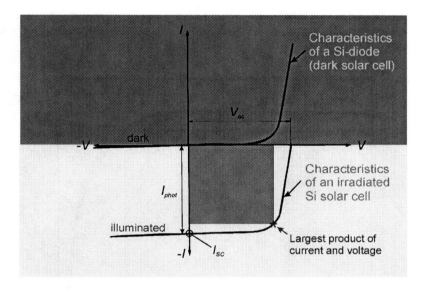

Fig. 2.3. Current-voltage characteristics of a diode (dark solar cell) and a irradiated solar cell with a short circuit current I_{sc} and open circuit voltage V_{oc}.

The short-circuit current I_{sc} is one of the most essential characteristics of a solar cell. It occurs in an illuminated, short-circuited solar cell. Open-circuit voltage V_{oc}: One describes the tension between the contacts if no current is taken (open circuit). The theoretically attainable (optimal) power which could be taken from the terminal, P_{opt}, is the product of short-circuit current I_{sc} and open-circuit voltage V_{oc}:

$$P_{opt} = I_{sc} \cdot V_{oc} \qquad (1)$$

---

[8] To measure the dark characteristic, an external variable power supply is necessary.

The attainable power P_{max} is defined by the greatest possible product of V and I at an operating point:

$$P_{max} = P_{mp} = I_{mp} \cdot V_{mp} \qquad (2)$$

The so called "Maximum Power Point" (MPP) is given by I_{mp} and V_{mp}.

The ratio of P_{max} to P_{opt} is called the fill factor FF. It describes the "rectangularness" of the trace of the characteristic. Figure 2.4 shows the power P as a function of the tension V and the corresponding I-V characteristic.

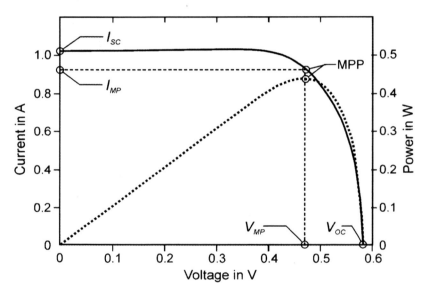

Fig. 2.4. Current-Voltage characteristics and Power-Voltage characteristics of a silicon solar cell. Also shown is the Maximum Power Point (MPP) at V_{mp} and I_{mp}.

The photovoltaic conversion efficiency η_{PV} is defined by the ratio of the electrical power output to the irradiated power on a solar cell. η_{PV} depends on irradiance and spectrum. The conversion efficiency is determined under standard test conditions (STC): an irradiance of 1,000 W/m² perpendicular onto the front surface, a cell temperature of 25°C and a spectral distribution according to solar irradiance passing at an elevation angle of 41.8° through the atmosphere (air mass 1.5). For physical reasons the photovoltaic conversion efficiency has a theoretical upper limit. This amounts to approximately 28% for crystalline silicon and has three main causes:

1. Silicon is a so called indirect semiconductor: This makes the absorption of a photon dependent on the occurrence of a phonon (lattice vibration) – while this happens relatively seldom, its absorption coefficient is comparatively low.

2. The band gap of silicon amounts to 1.1 eV: Photons with lower energy aren't absorbed at all, while photons with higher energy transfer the surplus energy to photons as lattice vibrations, i.e., as heat. This, and further losses, define the so called "spectral sensitivity" or "spectral response" of a solar cell (see below).

3. The maximum voltage (open-circuit voltage) V_{oc} depends on the difference of the potentials given by the p-n-transition and amounts to approximately 0.7 V for silicon.

This theoretical conversion efficiency is reduced in reality by different loss mechanisms:

- Optical losses, as reflection and shadowing losses caused by the front contacts and also losses by non-absorbed (transmitted) irradiance.

- Ohmic losses from series resistors (by contacts and sheet resistance) and by parasitic parallel resistors (see Figure 2.12).

- Recombination losses

The ability of a solar cell to convert an incoming photon of a specific wavelength into an electron-hole pair is called "quantum efficiency." The "internal quantum efficiency" neglects reflection losses on the surface of the solar cell, while the "external quantum efficiency" includes them.

While the energy of the photons (or "quanta" of energy) increases with their frequency (according to the law of Max Planck), each is usually creating one electron-hole-pair only with a constant energy potential. Therefore, spectral efficiency, defined by electrical energy output in relation to the irradiated energy, decreases as wavelengths become smaller. The spectral efficiency is best when the amount of energy from the incoming photon is just sufficient to create one electron-hole pair. If the energy of the photon is not sufficient to create an electron-hole pair, the photovoltaic effect is nil – this happens at wavelengths greater than 1,100 nm for silicon solar cells. Due to impurities in the silicon crystal the real spectral characteristic is somewhat different from the ideal described above, so it has to be measured. The so called "spectral sensitivity" or "spectral response" $S(\lambda)$ as shown in Fig. 2.5, 2.7, is defined by the photon density $j_{phot}(\lambda)$ divided by the intensity of incoming irradiation (irradiance) $E(\lambda)$ or $G(\lambda)$. Spectral efficiency is determined by measuring the response of a solar cell to modulation at a specific wavelength. The measurement is carried out by irradiating a bias spectrum (e.g., AM 1.5) in order to avoid disturbing effects caused by the irradiance level. Such a device is shown in Figure 2.6.

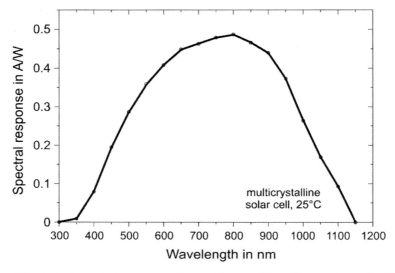

Fig. 2.5. Actual spectral response of a multi-crystalline silicon solar cell (ASE).

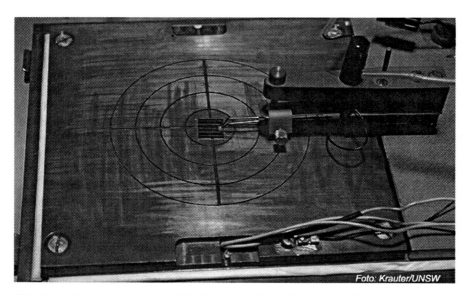

Fig. 2.6. Device to measure *I-V* characteristics, efficiency and spectral response. Shown with a high efficiency solar cell at UNSW.

The specific spectral distribution of the Sun's radiation on the Earth's surface is dependent on the thickness and condition of the atmosphere through which it has to pass. To some extend the efficiency and power output of solar cells are also dependent these conditions (see Figure 2.7, further information in chapter "irradiance modeling").

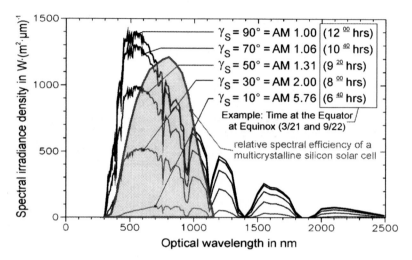

Fig. 2.7. Actual spectral response of a multi-crystalline (mc) silicon solar cell (ASE) together with the solar spectra for different elevation angles of the sun, equivalent air masses (AM), and time at the equator for Equinox.

2.3   Photovoltaic Generator

The internal electrical field in a solar cell is relatively weak and only small differences in electrical potential can be achieved (0.3 V for germanium and 0.7 V for silicon). The actual open circuit voltage that could be achieved is slightly less than these values. To get higher voltages, solar cells are connected in series – so called "strings" of solar cells.

Because these strings are very fragile, they are usually embedded in a soft plastic and glass sandwich creating a so called "solar module" or "PV module." The transparent soft plastic above and beneath the cell strings is usually the copolymer EVA (ethylene-vinyl-acetate), but PVB, silicones or TPU (thermoplastic polyurethane, a very recent development) are also possible options. In order to stiffen the compound and to make it more durable, a tempered front glass is added. If the backside consists of a composite foil (which consists of combination of layers such as PVF–aluminum–PVF or PVF–polyester–PVF), we call it a "laminated module"; if the backside is glass, we call it an "encapsulated module" (see Figure 2.8).

The output voltage at open circuit conditions of such a PV-module is usually set to 17 to 35 V for off-grid applications, which allows to fully charge a 12 V (respectively a 24 V) battery, so 36 to 72 cells are required in a series connection. The module is completed by attaching a small terminal box which contains the electric terminals and a metal or plastic frame which helps to mount the module and provides additional stiffness to it.

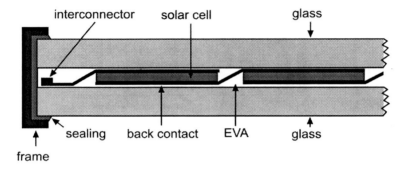

Fig. 2.8. Cross section of an encapsulated PV-module.

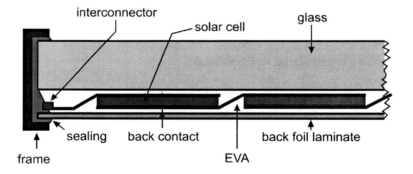

Fig. 2.9. Cross section of a laminated PV-module.

2.3.1 Electrical Properties

Electrical characteristics of a PV-module are indicated by some current-voltage characteristics, equivalent to the one of solar cells (see Figs. 2.3, 2.4). By connecting a variable ohmic load to the terminals of the irradiated PV module, combinations of current and voltage can be recorded which result in an *I-V*-curve when the load is varied.

Multiplication of the *I-V* pairs leads to output power which reaches its highest value at the so called "Maximum Power Point" (MPP). To make the data comparable, "Standard Test Conditions" (STC) have been set up: The spectrum is fixed and related to a sun spectrum at Air Mass 1.5, the irradiance is 1,000 W/m^2 and the cell operating temperature is set to 25°C. (IEC 904-1 and IEC 891, resp. DIN EN 60904-1 and DIN EN 60891, see also Annex). The variation of the irradiance E (or G) has only a small influence on the open-circuit voltage of a PV-module in a range of 350–1,000 W/m^2. At lower irradiance levels, the voltage decreases logarithmically (see Fig. 2.10). The short-circuit current is directly proportional to the irradiance because the current is equivalent to the number of electron-hole pairs generated by the absorbed photons. Consequently; the possible output power of a PV-generator is proportional to the irradiance from 350–1,000 W/m^2 (constant conversion efficiency).

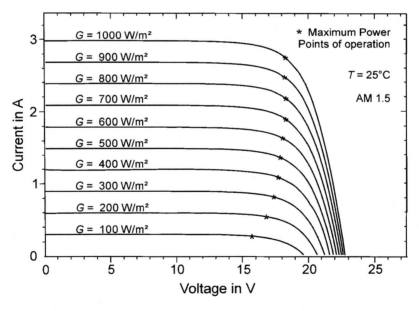

Fig. 2.10. Current-voltage characteristics of a multi-crystalline silicon PV-module for variations of irradiance (temperature and spectrum are kept constant at 25°C resp. AM 1.5)

For lower irradiance levels the conversion efficiency decreases due to voltage losses, which depend on the internal shunt resistance. Solar cells with a high shunt resistance are more suitable for low irradiance levels than cells with low shunt resistance (mainly caused by impurities in the cell material).

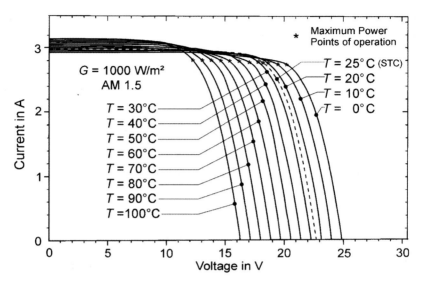

Fig. 2.11. Current-voltage characteristics of a multi-crystalline silicon PV module at variations of temperature (irradiance and spectrum kept constant at 1,000 W/m² resp. AM 1.5)

An increasing cell temperature at constant irradiance causes a reduction of the open-circuit voltage and consequently of the output power by -0.4%/K to -0.5%/K for crystalline silicon solar cells (see Fig. 2.12).

2.3.1.1 Equivalent Electrical Circuit

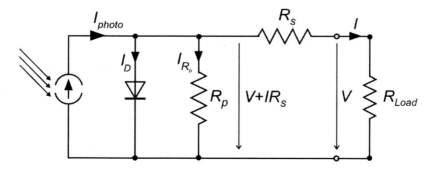

Fig. 2.12. Equivalent electrical network diagram of a solar cell according to the "one diode model".

The current-voltage characteristic of a solar cell approached by a "one diode model" (equivalent electrical network diagram see Fig. 2.12) could be described as follows:

$$I = I_{photo} - I_0 \left(\exp \frac{q(V+IR_S)}{kT} - 1 \right) - I_{R_p} \tag{3}$$

$$R_p = R_{p,dark} \cdot e^{-\alpha E} \tag{4}$$

$$I_{R_P} = \frac{V+IR_S}{R_p} \left(1 + a \left(1 - \frac{V+IR_S}{V_{br}} \right)^{-m} \right) \tag{5}$$

The current-voltage-characteristics of a PV module could be described as:

$$I = I_{photo} - I_0 \left(\exp \frac{q\left(\sum V + IR_S\right)}{kT} - 1 \right) - I_{R_p} \tag{6}$$

with:
a	avalanche factor	q	element. charge ($1.602 \cdot 10^{-19}$ As)
E	irradiance in W/m²	R_p	parallel (or shunt) resistor in Ω
I_0	diode saturation current in A	R_s	series resistor in Ω (0.05-0.5 Ω)
I_{photo}	photo current in A	T	absolute temperature in K
I_{Rp}	current of parallel resistor in A	V_{br}	breakdown voltage in V
k	Boltzmann const. ($1.381 \cdot 10^{-23}$ J/K)	α	coefficient for dependence of irradiance in R_p, in m²/W
m	avalanche exponent		

Usually, the parallel resistor is considered to stay constant. However, Zimmermann 1995 gives a typical $\alpha = 1.69 \cdot 10^{-3}$ m² W⁻¹, so the parallel resistor R_p is 350 Ω in the dark, at an irradiance of 1,000 W m⁻² it decreases to 70 Ω. Measured temperature coefficients for silicon PV modules could be found in Table 2.1.

2.3.1.2 Bypass Diodes

As with all series connections (e.g., at batteries or at PV), the element with the lowest current defines the total current. In order to avoid losses, only cells with an equivalent current at operation voltage are selected for series connection. Also the current of a cell may be reduced by local shadowing (e.g., due to dirt on the surface of the module) which therefore limits the total current and power output. If the string is large enough, the (reverse) voltage at the shadowed cell can surpass the negative breakthrough voltage and could lead to a local power dissipation that could even destroy the cell. To overcome this problem "bypass diodes" are switched in parallel (in opposite direction of the solar cell diode, often also called "antiparallel") to the solar cells or a small string of solar cells (see Figure 2.13). When a cell's current is reduced by shadowing, a reverse voltage builds up in the cell (assuming a load is connected) until it surpasses the breakthrough voltage of the bypass diode and thus part of the total current flows through the bypass diode, while the remaining I_{rest} passes through the cell.

Other ways of overcoming this problem are to use a low reverse breakthrough voltage of the solar cell diodes (e.g., at multi-crystalline solar cells) or a cell integrated bypass-diode (Green 1980: "Integrated Solar Cells and Shunting Diodes," Australian Patent 524,519; U.S. Patent 4,323,719).

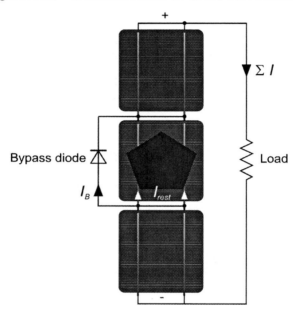

Fig. 2.13. Scheme of operation of a bypass diode: solar cell with partial shading in a string of 3 cells in series connection.

2.3.1.3 Electrical Terminals

The wires from the cell strings pass through the module laminate and are fixed by a pull relief. Then the wires are connected to an extension or end in a plastic box linked with a plugable or screwable terminal. Latest modules are pre-equipped with external cables with attached weatherproof plugs for module interconnection to reduce time for installation. Bypass diodes are integrated in the frame or are mounted in the terminal box.

2.3.1.4 Parallel Strings

As the efficiency of the power conditioning components (as inverters) increases with increasing voltage, modules are switched in series. In some countries (e.g., the US) safety regulation limits maximum voltage to 600 V (500 V incl. a security factor). For the most common size of cells (10 cm by 10 cm to 15 cm by 15 cm) this means a maximal power output of 2 to 3 kW_p.

To achieve higher power outputs without exceeding the maximum voltage of 600 V (or to get higher currents), modules (or strings of modules) are switched in parallel. In case of failure, such as voltage reduction in a string, e.g., caused by a higher temperature or by a shadowed cell (voltage is also lower when a bypass diode is applied), the remaining strings try to "feed" the defected string and may destroy it. Therefore, in order to protect the string, "string"-diodes are connected in series to each string of solar cells to avoid reverse currents (Fig. 2.14). If then the voltage of a string is reduced by a failure, the "string" diode is in blocking state and the remaining strings deliver their power to the desired load. A disadvantage of this configuration is permanent voltage loss at the string diode. Yet voltage loss can be reduced by using a low breakthrough voltage diode as germanium or Schottky barrier types [9]. Also, magnetic field sensors may be used to detect reverse currents and trigger an off-switch.

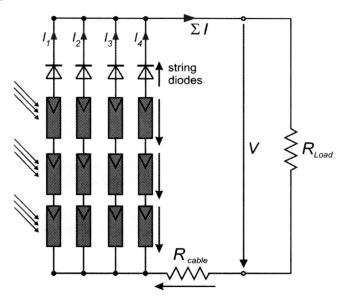

Fig. 2.14. Parallel connection of PV strings via string diodes for protection.

---

[9] Voltage loss of 0.3 V instead of 0.7 V at silicon diodes.

Another method is the use of relatively small "string"-inverters at the power conditioning side. The (synchronized) AC power output of each string is then paralleled.

2.3.2 Mechanical Properties

2.3.2.1 Sandwich Lamination

A lamination consisting of glass–plastic–solar cell–plastic–glass is called an "encapsulated" PV module, while a lamination of glass–plastic–solar cell–plastic is called a "laminated" PV module. The plastic is mainly a foil (thickness 0.5 to 0.7 mm) of EVA (ethylene-vinyl-acetate), which is processed at 150°C in a vacuum laminator: The copolymer EVA "cures" at that temperature and makes the lamination process non-reversible. The vacuum avoids air bubbles inside the laminate. To avoid degradation by ultraviolet sun radiation ("yellowing" or "browning"), a specific UV absorber is added to the resign. The foil on the backside used for the "laminated PV module" instead of glass is usually a compound of foils of Tedlar®–Polyester–Tedlar® or Tedlar®–Aluminum–Tedlar® at a thickness of 0.5 mm. Tedlar® is a polyvinyl fluoride film (PVF). The glass sheets of both types are made of iron free and thermal tempered glass at a thickness of 2 to 4 mm, to achieve a high optical transmission and to fulfill the ISO 203 regulations [10]. Further material properties are given in the Annex, Tables A10 to A14.

2.3.2.2 Framing

Frames of PV modules are mainly made of aluminum profiles, which hold the laminate onto the frame. The frame's corners are secured by stainless steel screws or saw tooth inlays. Small tube profiles made out of silicon rubber squeezed between laminate and aluminum framing, keep the laminate fixed, but however are tolerant to temperature expansion and to mechanical stress. Another, "cleaner" option is the use of a self-adhesive seal tape (e.g., Butyl) applied on the rim of the laminate, preceding the clenching of the frame. To reduce costs, but also for a better self-cleaning effect and a lower energy consumption for production, frameless modules are used in increasing numbers. The mounting of such modules to the structure is done by fixture compounds such as shown in Figure 2.15.

---

[10] Resistance for hail up to a diameter of 25 mm, torosion stability of the PV module for windspeeds up to 200 km/h.

2.3.2.3 Fixing and Mounting

Modules with frames are equipped with screw-threads or holes for fixing. Due to the required resistance to corrosion resistivity all screws and threads have to be made out of stainless steel in V2A to V4A quality (for maritime applications). For frameless PV modules, fixing is done by fixture compounds (Figures 2.15 and 2.16).

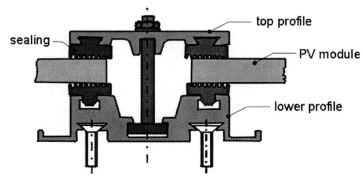

Fig. 2.15. Cross-section of a fixture-compound for frameless PV modules (after Schmid 1988).

Fig. 2.16. Roof-mounted PV generator with frameless modules and fixture compounds.

As well as the need for less material, handling is also advantageous since fixing and screwing is now at the front of the panel. A positive side effect is that less dust and dirt accumulation occurs at the edges of the module. Layers of dirt are washed away at rainfall, while framed modules tend to cause an accumulation of dirt which grows from the edges (at the frame boundary) to the center of the front surface of the module. Additionally, architects appreciate the more homogeneous visual appearance that is presented by the frameless PV modules.

A relatively new development toward cost-effective and visually appealing mounting of PV modules is the adhesion technology. First experiences, for example with the PHALK-Mont Soleil 560 kW_p in Switzerland, look very promising – the installation was faster and corrosion problems in the mounting compound consisting aluminum framing, stainless steel screws and steel support structures were eliminated. On the other hand, nondestructive dismantling of the module is not possible, yet in consideration of the extensive lifetime and reliability of PV modules, this factor is less critical. In the future rising costs of human labor and sinking module prices will make this approach even more favorable. However, while the thermal expansion coefficient of the plastics involved is larger than of that of glass, the module tends to be stressed considerably by bending during processing, so module size is limited.

2.4 Properties of PV Generators in Operation Conditions

The electrical power output of silicon solar cells decreases as cell temperature increases due to voltage-losses. For frequently used single- and multi-crystalline silicon solar cells, the voltage and power losses amount between 12% and 15% for an increase in temperature of 30 K (see Table 2.1). With conventional mounting or roof integration of the PV generators, cell temperatures of 30 K and more above ambient temperatures can be reached during a sunny day. Especially around noon, when irradiance reaches its peak value, the conversion efficiency of the solar cells becomes poorest due to the temperature effect.

Table 2.1. Measured temperature coefficients (TC) for Silicon PV modules

Type of PV module	$TC(V_{OC})$ in %/K	$TC(I_{SC})$ in %/K	$TC(FF)$ in %/K	$TC(P_{mp})$ in %/K
Single-Si # 1	-0.2817	0.0411	-0.1265	-0.3619
Single-Si # 2	-0.3413	0.0130	-0.1642	-0.5035
Multi-Si # 1	-0.2632	0.0435	-0.1172	-0.3318
Multi-Si # 2	-0.3675	0.0675	-0.1732	-0.4690
Multi-Si # 3	-0.2925	0.0407	-0.1556	-0.3996
ASE300-DG/50 (multi-Si)	-0.3726	0.1097	$TC(V_{mp})$ = -0.4752 $TC(I_{mp})$ = +0.0372	-0.4397
a-Si $_{min}$				-0.0393
a-Si $_{max}$				-0.2045

References: Emery et al. 1996; for ASE-300-DG/50: King et al. 1996

Until now the reference conditions used for the classification of PV modules (STC: Standard Test Conditions [11], SOC: Standard Operating Conditions) [12]

---

[11] STC: cell temperature 25°C, irradiance 1000 W/m² (perpendicular), sun spectrum equivalent to Air Mass 1.5 (see also IEC 60904-1, IEC 62145 and IEC 61215).

[12] SOC: as STC, but is using an actual measured cell temperature, occurring at an irradiance of 800 W/m², an ambient temperature of 20°C and a wind velocity of 1 m/s. Common values are between ca. 42°C to 57°C. SOC are achieving more

(continued...)

only give performances for one specific operating point (specific spectrum, perpendicular incidence, constant cell temperature and constant air speed). For the user, the knowledge of the yield in a certain period of time (including all occurring operation conditions) is more important. Therefore, examination and forecast of the actual daily temperature, efficiency profile and actual electricity generation is crucial, especially for the economic evaluation of a PV power plant. Considerable differences in estimations based on SOC, and especially of STC, could be observed. This is due not only to elevated temperatures, but also to optical reflection losses.

2.5 Mounting of PV Modules

While continuous R&D into more advanced production methods resulted in lower costs of solar cells and PV modules (see Fig. 2.17), expenses for installation and mounting remained constant or even increased due to elevated labour costs. The share of expenses for installation already amounts to 21%–53% (average 40%, see Wilk 1994, Strippel 1994).

Conventional mounting in the open field consists of a concrete foundation, metal tubes or profiles which are often even tailor-made to fit the size of the modules. Such a specific construction involves high materials and labor costs for as well as high maintenance costs due to corrosion susceptibility. Often, materials that require a lot of energy to produce such, as aluminum, lead to elevated energy-payback-times for the PV generating system.

In addition to improved module fixing methods such as the glueing technology (see chapter 2.2.2.3), which allows for cost reductions in mounting, innovations at the support-structure and the foundation could also occur (see chapters 10.2 to 10.4), resulting in faster installation by screw-less and foundation-less construction.

---

[12] (...continued)
realistic values for operation than STC (see also IEC 60891, IEC 61853 (draft) and IEC 61721).

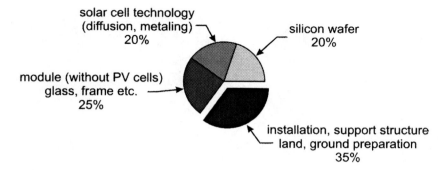

Fig. 2.17. Distribution of costs at a small PV power plant installation (without power conditioning, by Goetzberger 1994).

Currently, photovoltaics are most cost-effective in remote areas such as in alpine regions [13]. There, the transportation of materials and tools as steel bearers and concrete-mixer may cause difficulties however.

---

[13] In remote areas electrical grid connection is expensive, conventionally powered generators such as diesel generators require additional transportation costs for fuel, oil and spare parts. In these remote areas renewable energy generating sets often are the most cost-effective, even in todays economical conditions (Kayne 1992, Vallvé 1994).

40 Solar Electric Power Generation

Fig. 2.18. Conventional construction of concrete and aluminum for foundation, support structure and mounting (CEPEL, Rio de Janeiro).

Fig. 2.19. Conventional, expensive mounting construction in the Alps (from Wilk 1994).

2.6 Future Development of Photovoltaics

In previous decades, considerable advances have been made both in photovoltaic conversion efficiencies within the laboratory (Green 1995) and in reduction of production costs (see Fig. 2.22). Nevertheless, photovoltaic energy generation is competitive only in off-grid applications if social costs of fossil fuels are not considered.

Potential for development of single system components is not very far away from the theoretical optimum. While solar cells made out of materials other than silicon could achieve higher efficiencies, these materials do have low availability, are more expensive, and very often are less environmentally sound than silicon. There is a need for further development in the area of cost-effective production, composition of PV systems and installation (thin film technology, integrated power conditioning, applications). In order to maintain low production costs, details applied to laboratory samples could not be transferred into mass production. As a result, solar cells available on the market reach maximum conversion efficiencies of 20% to 21% (e.g., Sunpower), while laboratory samples reach up to 25% (e.g., UNSW, ISE). Due to necessary series connection of cells additional ohmic losses occur and current possible is limited by the weakest link, the worst cell. Additionally, the whole module area cannot be covered by solar cells due to space requirements for electrical insulation and thermal expansion of the cells. This necessity limits the module area that can be covered by solar cells and consequently the overall efficiency reaches only 15% to 17%. Under real operating conditions the following additional losses could be observed:

- optical reflection losses due to non-perpendicular irradiance (Krauter 1994a),

- losses due to low irradiance levels (reduction of form factor and voltage)

- thermal losses as voltage reduction due to elevated cell temperatures (Krauter 1993c)

- Reduction of output current for irradiance sun spectra with an air mass lower than AM 1.5.

- Shadowing: If a cell is shadowed in a serial string, the output current is limited by the reduced current of the shadowed cell. Bypass diodes can avoid this effect to some extent. The leakage current of bypass diodes may provoke some performance losses also. If parallel strings are used in the solar generator, strings with low voltage due to shadowing would operate as load. Therefore, serial diodes are used to avoid load condition on the string. On the other hand voltage losses of 0.3-0.7 V (depending on the type of diode) at the serial diodes occur.

- Power conditioning units are very often located in a small building some distance away from the generator. According to the literature, the wiring losses from the generator to the converters are in the vicinity of 3% for most applications.

- The inverters often have a high conversion efficiency at the rated power input, but for low irradiance levels and low power input the conversion efficiency decreases. Therefore, the average conversion efficiency over a whole day could be considerably lower than the rated one.

In considering all these losses, the relation of irradiance to generated electrical energy over a year is only 10:1 to 12:1. To take account of these losses and the consequent differences with the expected yields, literature uses a vague correction factor, "Performance Ratio" (accuracy 10-20%), to makeup the difference between the expected power output rating and real performance. Recently, this expression is also used for inverters. Detailed analyses on the causes of performance losses are rare, as are constructive improvements to avoid them. Some approaches to compensate these deficits are presented in this work.

Under real operating conditions the yields achieved are 23-45% lower than the ones in the laboratories or under Standard Test Conditions (STC). Investigations of that effect have been carried out within I-MAP (Intensive Monitoring and Analysis Program) as part of the German "1,000-Roofs-PV-Program" by Kiefer 1994 and at the PV power plant "Neurather See" by Voerman 1984.

By improving the optical and thermal properties of PV module installation, a gain of electrical yield of 12% could be achieved under real operating conditions, which brings operational efficiency much closer to nominal efficiency (Krauter 1993c, 1994b, 1996a).

As a result of greater efficiencies and lower costs of PV system components (solar modules and inverters), the importance of installation and mounting issues is growing. Costs, especially for mounting the module, are disproportionally high, as could be seen in Fig. 2.17. However, there is a huge potential to reduce the costs of these cost sectors and ultimately lead to a faster spread of photovoltaics in the energy market.

By expanding the knowledge of the interface of the PV module and real environment, efficiency gains could be expected that exceed the ones achievable in solar cell technology.

2.7 Research Funding for Photovoltaics

Public funding on research of terrestrial photovoltaics started after the 1973 oil crises. Figure 2.20 charts funding from 1974 to 1996 (resp. to 1992 in the US). Clearly visible, until 1980, is the intense funding by the U.S. Carter Administration and the significant decrease by Reagan. In Germany a drop in direct funding during 1990 to 1994 could be observed in favor of indirect funding within the "1,000-PV-Roofs-Program" (which then actually supported 2,250 PV roofs). During that time the direct funding budget in the US increased constantly.

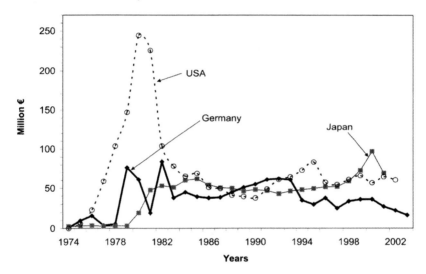

Fig. 2.20. Direct governmental funding for PV R&D in Japan, USA and Germany. References: Sissine 1994, Sandter 1993 resp. Brauch 1997, US Department of Energy; and International Energy Agency (IEA-PVPS), based on 2003 US$ exchange rates. Exchange rates between US Dollars, German Marks resp. Euros are available in Table A 16 in the Annex.

Funding for photovoltaics and other renewable energies is at a factor of eight below the equivalent expenses for the research of nuclear energy. Only 9.2% of the German energy R&D budget was spent on solar energy in the years 1981 to 1990, while for example Denmark was spending 26.9% of the equivalent budget on renewable energies (Scheer 1993).

In Germany a hundred times the renewable energy budget is spent to support uncompetitive coal mining (in 1997: 4.57 billion €; in 1998: 4.74 billion € – see Table A6). The subvention for each miner and other personnel of the mining industry amounts to 51,000 €/a. This would be sufficient to create twice the number of jobs in the solar energy sector.

2.8 Market Development of Photovoltaics

Development of PV production worldwide is shown in Figure 2.21. Production of PV modules in terms of output power under Standard Test Conditions (indicated by the index "p," deriving from "peak") increased from 3.3 MW_p in 1980 to 1,000 MW_p in 2004. This corresponds to an average yearly growth of ca. 14%. The total installed power at the end of 2004 was 4,000 MW_p. This development appears quite satisfactory, but in comparison to conventional power plant capacity the value is still small.

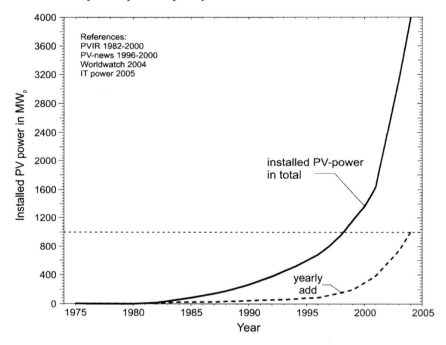

Fig. 2.21. Market development of photovoltaics: Installed capacity and yearly production (References: PV-News, PVIR (Photovoltaic Insiders Report) 1982–2000, Worldwatch Institute 2004, IT Power 2005).

In 1973, prices for PV modules were about 700 $US/$W_p$ and consequently were only used for aerospace applications. Terrestrial PV modules had been offered for 70–80 $US/$W_p$ in 1975, nowadays (2000) terrestrial modules are already available for 2–3 $US/$W_p$. Development of prices is plotted in Figure 2.22. An example of the current production costs is given in Table 2.2.

Relative cost shares for complete PV installation which consists of PV modules, installation, support structure, land, and ground preparation (without power conditioning and energy storage) are shown in Figure 2.17.

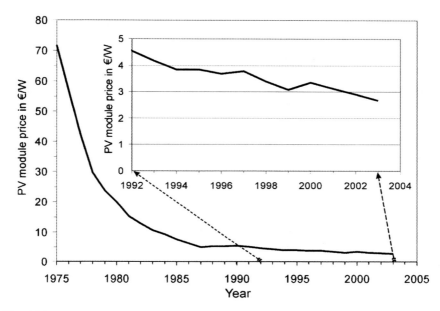

Fig. 2.22. Development of prices of photovoltaic modules (in € per W_p) from 1975 to 2003 with an amplification of the recent period from 1992 to 2003 (data given in US$ has been referenced to 1998). Source: Worldwatch Institute 2003, IEA-PVPS 2005.

Terrestrial world market shares in PV systems in the early 1990's:

- Communication (e.g., transmitters) — 21 %
- Solar-Home-Systems (1-2 modules) — 16 %
- High-power consumer applications — 15 %
- Water pumping systems — 11 %
- Grid-connected systems — 11 %
- Remote living (> 2 modules) — 7 %
- Small power consumer applications (watches etc.) — 6 %
- Village power supply — 4 %
- Cathodic corrosion prevention — 3 %
- Military; signalization — 3 %
- Remote (other) — 3 %

During the last decade the grid-connected share grew from 11% to more than 70% in 2002. This development was triggered by governmental support programs in the first world, while governmental support for off-grid PV systems, mainly located in developing countries, was less, numbers of off-grid installations increased steadily, but their relative market share decreased. Within the off-grid sector the relative share of the different applications remained the same.

Table 2.2. Production costs of multi-crystalline silicon PV modules in Germany

Reference	Mertens 1992 and Staiß 1995 in DM/W_p			LBST 1995 in DM/W_p		Frantzi et al. 2000 in $/$W_p$	
Process	lowest prices	average price	highest price	5 MW/a	20 MW/a	in 2000 10 MW/a	in 2010 100 MW/a
Silicon (basic material)	0.51	0.58	0.50				
Casting	0.82	1.11	1.37	2.25	1.79	0.82	-
Wafer production	0.80	1.13	1.00				
Cell fabrication	0.70	1.48	2.38	2.46	1.32	0.44	-
Production of module	1.03	1.85	2.62	1.28	0.87	0.86	-
Common costs, profit	-	-	-	1.80	1.20	-	-
Total	3.90	6.20	7.80	7.79	5.18	2.11	1.15

Areas of application for autonomous PV-Systems in comparison to conventional energy supply technologies are presented in Figure 2.23.

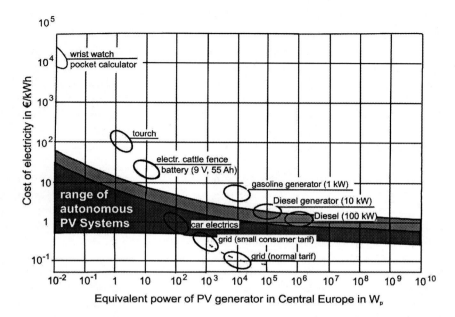

Fig. 2.23. Costs of autonomous PV applications as a function peak power of PV generator and compared with other methods of electrical energy supply (Schmid 1995). Light grey is for Central European conditions (1,000 kWh a^{-1} m^{-2}), dark is for locations closer to the Equator (2,000 kWh a^{-1} m^{-2}).

Distribution of Cell Production by Technology

(Reference: P.D. Maycock, PV market update, 2003)

- Multi-crystalline Silicon: 58%
- Single-crystalline Silicon: 32%
- Thin-film technology: 7%
- Others: 3%

Governmental Policies for Market Introduction of PV - Germany as an Example

To accelerate market penetration of renewable energies, the German Government is paying a sustainable fee for each kWh of photovoltaic electricity injected into the public grid since April 2000. The payment of renewable electrical energy generation was already very successful in the field of wind converters which led to increments of installed power by 30% per year over the last decade (the total installed wind power generation capacity in Germany reached 16,629 MW in 2004, by far world's largest capacity).

While formerly entire PV-installations been financed, the technology advanced (as for grid inverters during the "1,000 PV Rooftops Solar Electricity Program") but the prices of PV-system components remained stable or even rose slightly due to a stable demand (as for PV modules). By just paying for the energy generated, the plant operator is now forced to set up and to keep his installation as effective as possible. Monitoring the yield of the power plant over the entire lifetime of 25–30 years is now done by the home owner himself. Competition of different manufacturers led to cost and technology improvements. From January 1999 until the end of 2003 the "100,000 Rooftops Solar Electricity Program" was carried out, with a granted capacity of 345.5 MW and 65,700 systems built.

Before that programs mentioned above, some cities and federal governments, forced by a lot of solar initiatives, paid up to 1.1 € for each kWh grid injected photovoltaic electricity. The general "Electricity Feed Law" introduced in 1991 was replaced by the "Renewable Energy Sources Act (EEG)" in April 2000 by the Federal Government. In 2004 improved feed-in tariffs ranging from 0.454 to 0.574 €/kWh, depending on size and installation type (ground installation, roof mounted installation above 30 kW_p and roof mounted systems below 30 kW_p) without limits for the total number of installations. This regulation boosted the installation and production in Germany to 300 MW_p per year. Within a short time Germany became a world leader in terms of installed PV capacity per year, is and is very likely to overtake the actual PV production leader Japan within 2006. In order to stimulate progress also for cost reduction measures, the PV compensation is reduced gradually by 5% per year.

3
Inverters

3.1 Autonomous Operation

While electrical power generation changes due to different irradiance levels, the energy still has to be stored, especially when the time of demand is different from the time of generation - which is very often the case. Therefore storage consisting of a battery and an adequate charge/discharge controller has to be added to the system (see Figure 3.1). Often alternating voltage is needed, and thus an inverter has to be implemented in order to convert DC from the PV panel and storage to the AC required.

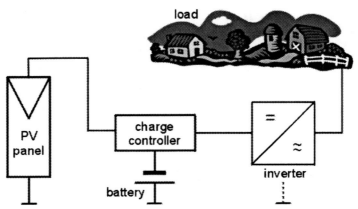

Fig. 3.1. Block diagram of an autonomous PV system with storage and an inverter for AC loads.

The most simple way to achieve this is by switching the polarity of the DC with the frequency needed for AC (50 Hz or 60 Hz), using devices called rectangular inverters. Yet his kind of AC conversion leads to high distortion levels with higher frequency artifacts, which could damage sensitive loads and interfere with radio signals.

A better approximation of the sine-form requested is to use switches with some zero voltage period (so called trapezoid inverters). Distortion levels for this type of inverter are lower than for the rectangular inverters, but still high.

Formerly so called "rotating inverters" have been used. At those, a DC motor is coupled to an AC generator. This method enables a very smooth sinusoidal output, but efficiency is relatively poor and the frequency varies when a change of the load occurs.

State-of-the-art technology at inverters is PWM (pulse width modulation). Here DC is switched on and off for a short period of time (a pulse) in such a manner that the integral of the pulse is equivalent to the actual level at the sinus profile required. The next pulse width is also adapted in that way so the integral is equivalent to the next actual level of the sinus given by the controller. After filtering the output of such an inverter is very close to a perfect sinus at relatively high efficiency (up to 96%).

3.2 Inverters for Electrical Grid Injection

In autonomous PV systems the energy yield varies due to daily and seasonal changes of solar irradiance. In Central Europe the average irradiance received during summer is about five to six times higher than in winter. Therefore those PV systems have to be equipped with enough energy storage devices to supply loads during periods of poor or no radiation. Storage makes a system more expensive (especially seasonal storage) and increases the costs of generated energy. Therefore, the public grid is used as "storage" or buffer into which energy is fed during periods of overproduction and likewise, taken during periods lacking photovoltaic power generation [14]. To facilitate grid injection, DC power from the PV generator has to be converted into AC according to the requirements of the public grid.

Counter to inverters of autonomous systems, the reference sinus for the PWM is not given by the controller but by the grid. While the impedance of the grid is generally very low, the inverters still have to be synchronized before operation and adapt to the particular voltage and frequency of any given grid. Therefore, inverters used for grid connections are different from the ones used for autonomous systems. In case of any connection to the grid, utilities set standards in order to maintain distortion levels and synchronization of voltage and current within certain limits. In the past numerous failures had been reported which involved grid connected inverters, the technology nowadays has been found to be very stable. Only a

---

[14] According to different examinations the public grid in many countries (as Germany) is capable of allowing PV grid injection in a vicinity of at least 15% of its nominal grid power.

few types of inverters are capable of working in both operation modes: grid-connected and autonomous.

While small loads can often be used for DC, general appliances require AC of 115 V or 230 V with 50 Hz or 60 Hz. For power requirements higher than 3-10 kW a three-phase AC system is usually preferred.

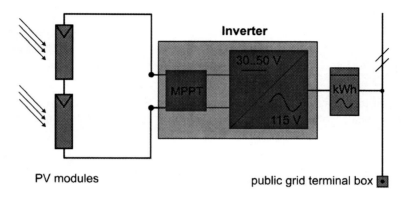

Fig. 3.2. Scheme of a typical single-phase PV grid injection system equipped with a Maximum-Power-Point-Tracker (MPPT) and an energy counter (kWh).

Many different types of PV grid injectors are available. They differ by their way of commutation whether they feature galvanic separation from the grid (e.g., by transformers), and by the type of power electronics they are using (thyristor, GTO, bipolar-transistor, MOS-FET or IGBT). Due to adequate funding programs for PV power plants of these dimensions (e.g., the German 1,000 and 100,000 PV Roofs Program), most of these grid injectors are available in a power range of 0.6 to 5 kW.

A widely utilized principle is the pulse-width-modulation (PWM). For PWM to achieve a 50 or 60 Hz sinus for an AC grid, the first step is to get a rectangular current by shifting the polarity (each 10 ms for 50 Hz, or each 8.33 ms for 60 Hz) of the DC output from the PV power station. The rectangular current is then switched on and off ("pulsed") in such a manner that the resulting integral of the pulses is as close as possible to the equivalent sinus value to be achieved (see Figure 3.3). The pulsing is carried out at a relatively high frequency (10 to 100 kHz), the consequent integration is done by "smoothing" the pulses via a lowpass filter.

The frequency and phase have to be adjusted to the actual grid condition and if the "guiding" grid fails the inverter also has to be switched off immediately. In general, a grid inverter should fulfill the following requirements:

- The output current follows the grid voltage synchronously ("current-source").
- The distortion and consequent spectral harmonics of the grid frequency are not allowed to surpass given thresholds by the norms (e.g., VDE 0838, EN 60555), which requires a good adaption of the output current to the sinus form.
- The injected current and the grid voltage should have no phase difference ($\cos \rho = 1$) in order to avoid the bouncing of reactive power between grid and inverter which would cause additional losses and eventually overcharge.
- In failure condition (missing grid voltage, strong frequency shifts, short circuits or isolation failures) the grid injector has to be disconnected from the grid automatically.
- Control signals in the grid, often used by the energy suppliers, should not be disturbed by the grid injector nor should they impact his operation.
- The input terminal resistance should be well adapted to the actual properties of the solar generator during operation, e.g., by "Maximum Power Point Tracking" (MPPT).
- Fluctuations of the input voltage (e.g., at 100 Hz caused for a single phase injector device) should be low (< 3%) to allow operation of the solar generator close to its Maximum-Power-Point.
- Overvoltage, e.g., caused by a solar generator operating at low temperatures near open circuit conditions, should not lead to defects.
- For overload conditions the input power is limited to a defined value by shifting the point of operation of the PV generator toward open circuit voltage. This could happen when the nominal power of the inverter is lower than the nominal power from the PV generator. Such an occurrence infrequently (see Figure 7.27) leads, at very high irradiance, to an input overload of the grid injector.
- The grid inverter should be supplied by the solar generator to avoid consumption from the grid (e.g., at nighttime). The inverter should go into operation mode already at low irradiance levels and should operate in a stable manner. Modern PV systems are already injecting to the grid at irradiance levels of 50 W/m² and efficiencies of 90% can be achieved at 10% of the nominal inverter power so far.
- Input and output terminals should be protected against transient overvoltages ("surge", e.g., caused by lightning strokes). This is mainly carried out by the application of overvoltage or surge arrester devices.

- The regulations for "electromagnetic compatibility" (EMC), e.g., EN 55014, have to be adhering.
- Noise emission of the devices should be low, so as to allow operation also in inhabited buildings.

The quality of power entering the utility-grid from a PV system is also of concern to the utilities. If too many harmonics are present in the inverter output they may cause interference in loads at other locations (that may require sinusoidal power) or at utility equipment (e.g., for data transmission over the transmission line). Electrical machinery (e.g., motors) operating with a lot of harmonic distortions in the power supply are heating up and the lifetimes of the bearings are reduced due to vibrations. Regulations are given in IEEE 929 and IEEE 519 which set the limits for harmonics as shown in Table 3.1a.

Table 3.1a. Limitations for harmonic distortion for grid connected PV systems

Odd harmonics	Distortion limitation
3rd through 9th	4.0 %
11th through 15th	2.0 %
17th through 21st	1.5 %
23rd through 33rd	0.6 %
above 33rd	0.3 %

Even harmonics should be less than 25% of the odd harmonics in the ranges listed. References: IEEE 929, IEEE 519-1992, Messenger et al. 2000

IEEE 519 is specific to grid-connected PV systems and also gives standards relating to:

Voltage disturbances: Voltage at inverter output should not be more than 5% higher than the voltage at the point of utility connection and thus the inverter should sense grid voltage abnormalities and disconnect the inverter when indicated. Disconnection should occur within 10 cycles if the utility voltage either drops below 50% of its nominal value or increases above 110% of its nominal value. If the utility line voltage is between 50% and 92% of its nominal value, the inverter should shut down within two seconds.

Frequency disturbances: If, at 60 Hz systems, the line frequency falls below 59.5 Hz or goes above 60.5 Hz, the inverter should be disconnected.

Power factors: The power factor (caused by a phase shift between current and voltage) should not be lower than 0.85.

Injection of DC into the AC grid: DC current must be no greater than 0.5% of rated inverter output current.

Also, regulations on islanding protection, reconnecting after grid failure and restoration, grounding surge protection, DC and AC disconnecting are assumed in that norm. Sometimes additional requirements of the local energy utilities have to be taken into account. A collection of further international and European regulations are given in the Annex.

A typical configuration of a single phase PV injection system is shown in Figure 3.2 (with a maximum-power-point-tracker and an energy counter).

3.3 Types of inverters

3.3.1 External Commutated Inverters

External commutated inverters need an external AC voltage supply (which is not part of the inverter), to supply the "commutation-voltage" during the commutation period (e.g., for thyristors, see also DIN 41750 part 2). At grid controlled inverters this AC voltage is supplied by the grid. External controlled inverters are operated via "natural commutation." A main feature of them is that a "current rectifier valve" with an actual higher voltage potential after ignition takes over the current from a current rectifier valve in front of it (Heumann 1996).

The external commutated, grid-controlled inverter is commonly used for high power applications. For low power applications (<1 MW), which are most common for PV power supply systems, self-commuted inverters are the state-of-the-art.

3.3.2 Self Commutated Inverters

Self-commutated inverters do not need an external AC-voltage supply for commutation (see DIN 41 750, part 5). The commutation voltage is supplied either by an energy storage which is part of the inverter (commonly by a "delete"-capacity) or by increase of resistance of the current rectifier valve to be switched off (e.g., a MOS-FET power transistor or an IGBT). Self-commutated inverters have been designed for all kind of conversion of electrical energy for energy flows in one or both directions. In the power range relevant for PV-applications (< 1 MW) nowadays exclusively self-commutated inverters are used.

3.3.3 Inverters Based on PWM

A self-commuted inverter, which has an output voltage (resp. current) that is controlled by pulses is called pulse-inverter. At this type of inverter the number of commutations per period is increased by frequent on- and off-switching at the pulse-frequency f_p within this period, which may be used to reduce harmonics of current and voltage, because it is equivalent to an increment of pulse numbers. At grid-controlled inverters the increment of pulse numbers is only possible via an adequate augmentation of inverter rectification braces. Figure 3.3 shows the coupling of a DC voltage source (PV-generator) with an AC voltage source via a pulse inverter in a single phase bridge circuit.

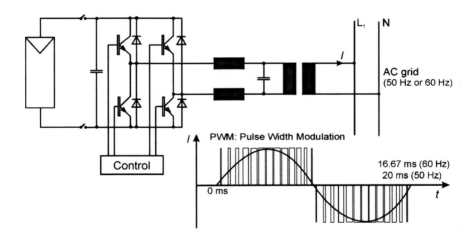

Fig. 3.3. Scheme of circuit of a single phase sinus inverter based on pulse width modulation for photovoltaic grid injection.

The harmonics of the current are defined by the inductances of the AC side. So, in order to fulfill the directives of the utilities for grid-feeding (EN 60 555), a certain minimum inductance should be kept.

The inverter shown in Fig. 3.3 supplementary is equipped with a low pass filter and an isolation transformer, so all harmonics up to the order of $n-1$ are eliminated, while n represents the number of pulses during each period of the AC current. For elevated switching frequencies the switching losses in the power electronic devices are increased. At low switching frequencies the expenses for the low pass filter increases. While for the sinusoidal current the power into the single-phase grid pulses with twice the frequency, the DC from the PV-generator is superimposed by a sinusoidal current with twice the grid frequency.

3.4 Electrical Grid Connection

3.4.1 Voltage Levels of Electrical Grids

Voltage levels for grid-feeding into public electrical grids (recommendations according to VDE/IEC, in square brackets: the actual practice in Europe):

Low voltage:	230/400 V	Supply of small consumers in households, agriculture and industry
Medium voltage:	10 kV [12 kV]	Public and industry grid systems
High voltage:	110 kV [123 kV] 220 kV [245 kV] 380 kV [420 kV]	For countrywide grid system depending on space and capacity
Very high voltage:	756 kV	Countrywide grid systems, transmission of high power over large distances

So far, PV grid-feeding has been carried out in the low- and medium voltage range only.

3.4.2 Boundary Values of Electrical Grids

The inverters applied have to fulfill the requirements set up by the utilities (mainly concerning the maximum of harmonics) as well to withstand the possible impacts by over voltage (surge) of the grid on the inverter.

Short- and long-term voltage tolerances are prescribed for AC-grids that are designated for the connection of inverters. The bankable deviation from the ideal sinusoidal shape is regulated as well. Also the impact of the inverter operation on the shape of the grid voltage, in particular during the time of inverter commutation, has to be taken into consideration. For the long-term tolerance of the grid voltage a.fluctuation of the root mean square value of the AC-voltage between 90% and 110% of the nominal grid voltage is permitted (according to VDE 160, part two and international norms). Besides long-term voltage fluctuations, also short-term, non-periodic overvoltages may occur; their size is conditioned by the temporal development. Their value can range from 1.5–2 times of the nominal voltage for discharged (released, unloaded) generators up to 10–100 times for statics. Elevated surges have to be limited to a maximum of 2.5 times of the nominal voltage by the applying adequate surge-arresters at least in high- and medium voltage grids.

3.4.3 Long-Distance Transport of Electricity

For distances greater than 1,500 km, high voltage DC transmission (HVDC) is generally more advantageous than transmission by AC. Additionally, it offers the following features: Lower peak voltages, so insulators and safety distances could be smaller, leading to fewer environmental impacts (landscape-aisles) for the same power transmitted. Also, there are no reactance- and EMS (EMV) problems. HVDC technology has been applied for 25 years within electrical energy technology and can be considered ripened. Altogether transmission lines of 11,000 km are in operation, and 4,500 km more are under construction.

For example an electrical power of 6,300 MW has been transmitted by a HVDC system of 1.2 MV (± 600 kV) through Brazil since 1994, covering a length of 805 km from the 12.6 GW hydropower plant Itaipu at Foz de Iguaçu to the São Paulo region (Ibiuna). The 18 kV, 50 Hz, three phase AC from the generator is transformed to 525 kV in the first stage and then rectified in the rectifier station at Foz de Aegis. After the long-distance transport by a 600 kV positive and a 600 kV negative line, the HVDC is inverted in Ibiuna to the customary 60 Hz [15] and is finally injected into the public electrical network of São Paulo, the second largest city in the world. In the IKARUS [16] project the import of solar electricity to Germany was used as a reference to describe a HVDC transmission line of 2 GW at 400 kV (Table 3.1).

---

[15] The use of HVDC offers an additional advantage for this case: Aside from the transport, the frequency of 50 Hz by nine of the eighteen 715 MW generating units is transformed to 60 Hz. For political reasons - the power plant is located on the borderline between Brazil (public grid of 60 Hz) and Paraguay (public grid of 50 Hz) - half of the generators had to be built in 50 Hz, nevertheless Paraguay is just using 2% of the generated energy, while Brazil is using all the rest.

[16] IKARUS: Instruments for the development of strategies to reduce greenhouse gas emissions caused by energy use. Project by the German Ministry of Science and Technology, terminated in March 1995.

For the connection of a PV power plant in Southern Spain to the consumers in Germany the length of the HVDC transmission line between the terminal stations is 2,000 km. For the connection from Northern Africa to Germany a length of 1,200 km of transmission lines on the African continent was assumed to reach regions with very high solar irradiance. From Tunisia a 200 km long sea-cable is linked to Sicily, then another 1,900 km overland transmission finally brings the electricity to Germany. The average yearly efficiency is 88% for the connection from Spain and 84% for the connection from Africa. This means that the transport losses are between 12% to 16%, equivalent to 0.5% per 100 km of transmission and 1.5% for the terminal stations. The costs for the terminal stations are 84.4 €/kW, the overland transmission lines cost 30,675 €/km and the cable 1.02 million €/km. Thus the total investment costs are a little less than 1.02 billion € for the connection from southern Spain to Germany, respectively 1.53 billion € for the connection from Northern Africa – which are only about 30% of the costs for the yearly subsidy to carbon mining in Western Germany (see Table A6 in the Annex).

Table 3.1b. HVDC Transmission to Central Europe from different locations

Location	Southern Spain	North Africa	Jamal	NE Africa
Power (total)	2,000 MW	2,000 MW	26,500 MW	72,414 MW
Power (Net)	1,760 MW	1,679 MW	24,644 MW	67,055 MW
Voltage	400 kV	400 kV		
Current	2 x 2.5 kA	2 x 2.5 kA		
Transmission line (overland)	2,000 km	3,100 km		
Transmission line (sea cable)	0 km	200 km	4,100 km	5,100 km
Efficiency during a year	88%	84%	93.1%	92.6%
Costs of transmission lines (overland)	0.62 billion €	0.95 billion €	7.8 billion € (sea & land)	24.6 billion € (sea & land)
Costs sea cable	–	0.20 billion €		
Costs of transmission lines (sea)	2 x 0.17 billion €	2 x 0.17 billion €	1.63 billion € 1.52 billion €	4.46 billion € 4.13 billion €
Operational costs	3 mio €/a	4.3 billion €/a	111 mio €/a	262 mio €/a
Transmission costs	20–60 €/MWh	25–77 €/MWh	3.0 €/MWh	3.8 €/MWh
References	Staiß 1996		Czisch, G., ISET, 1999	
	Lifetime of transmission line: 30 years, operational personnel: 40 respectively 49 persons; transmission cost at an interest rate of 4%/a.		Lifetime of transmission line: 25 years, interest rate: 5%/a	

Utilizing current technology, total costs of imported PV electricity delivered via the public grid in Germany using technology available today are 0.2 €/kWh to 0.34 €/kWh. Compared to local PV electrical generation in Germany, these costs are less than half. Importation of solar electricity offers additional advantages: Short-term fluctuation of electrical generation due to irradiance is less common, and while total irradiance is higher, seasonal changes of irradiance are less consequential.

4
Storage

Often, electrical energy is needed in during other periods of time than it is generated. In the case that an electrical grid connection is not possible, the energy has to be stored in the period between generation and consumption. By principle many different kinds of storage are possible:

A. Electrical storage
A.1 Capacitors (expensive, self-discharge is relatively high, recently used for short-term storage, e.g., "Gold-Caps®," "Ultra-Caps®"),
A.2 Inductors (little storage capacity, high voltage, practically not used)

B. Mechanical storage
B.1 Kinetic energy (flywheel, sometimes used for mid-size, short-term power storage)
B.2 Potential energy (water storage in high altitudes, e.g., mountain lakes, pumps in combination with hydro-electrical power generation)

C. Chemical storage
C.1 Electrolysis (electrolysis of water to hydrogen and oxygen as storage medium, later burning or transformation into electricity by fuel cells)
C.2 Electrochemically (chemical transformation of the electrolyte and electrodes are generating an electrical potential (first discovered by Alessandro Volta that where the name "Voltage" derives from). Reversing this process ("charging") is carried out by applying a voltage to the electrodes).

The most common way of storing electricity for photovoltaic energy systems is the electrochemical storage by the lead battery with sulphuric acid as an electrolyte. Therefore, this method will be discussed more in detail. The other processes are not very common and therefore are just mentioned in principle.

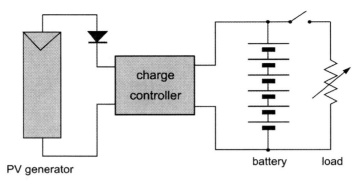

Fig. 4.1. Scheme of a PV system with a chemical battery storage.

4.1 Lead Sulphide Acid Battery

Lead sulphide acid batteries, also called "lead-acid" batteries, invented in 1859 by the French physicist Gaston Planté are the most commonly used rechargeable batteries today, at least for high capacity requirements. While the energy-to-volume ratio is acceptable, the energy-to weight ratio is rather low compared to other battery types. They are relatively cheap and can supply high surge currents. Each cell has a nominal voltage of 2 V, which can be increased by a series connection of battery cells.

4.1.1 Principle

Two electrode plates, one consisting of pure lead and one with a lead oxide surface are in an electrolyte consisting of thinned sulfuric acid (typically 37% H_2SO_4). During discharging both electrodes turn into lead sulfate ($PbSO_4$) and the electrolyte turns into water, therefore discharged lead-acid batteries can freeze. During charging, the lead sulfate on both electrodes is converted back to lead dioxide (PbO_2 at positive electrode) and sponge lead (Pb at negative electrode), and the sulfate ions (SO_4^{2-}) are driven back into the electrolyte solution to form sulfuric acid (see equation of chemical reaction below).

$$PbO_2 + Pb + 2\,H_2SO_4 \Leftrightarrow 2\,PbSO_4 + 2\,H_2O$$

4.1.2 Gassing

Gassing occurs when more current is being fed to the battery than it can use. The excess current produces Hydrogen and Oxygen gasses. Some gassing is normal, but excessive gassing can indicate that the batteries are being overcharged. The gasses released are explosive if exposed to a spark or flame, therefore adequate ventilation must be provided. Batteries normally start gassing at about 80-90% of full charge. A common fallacy is to stop charging as soon as the battery starts gassing. Most batteries start gassing at about 80% SOC, so if the charge is quit at the point of gassing, the battery will never reach full charge. To prevent excess gassing most better chargers cut back on the current when the battery reaches this point.

Gassing at sealed gel cells: Gelled cells will withstand much less heavy gassing than flooded batteries. The gel can develop large bubbles or "pockets," which reduce battery capacity due to poor contact with the plates. It can also cause the gel to dry out from water loss, making these pockets permanent. Gelled batteries are charged at a slightly lower voltage than flooded batteries, 0.1 to 0.3 V less to avoid over-gassing. Some advertisements and product brochures have stated that gelled cells have a "high" capacity for taking a charge – this is not correct, as it is ½ to ¼ the maximum current that a flooded battery can take.

4.1.3 Specific Gravity (SG)

The measurement used to express electrolyte strength. Specific gravity (SG) compares the weight of the electrolyte to water (SG of 1.000 kg/l). SG changes with temperature so most hydrometers come with a correction chart. A full charge should be about 1.265 at 25°C, yet changes with temperature. SG cannot be measured in sealed batteries. Pure acid has a SG of 1.835 kg/l. A fully discharged battery will have a SG of about 1.120 kg/l. SG should not be measured right after water is added as the reading will not be accurate until the electrolyte is fully mixed. This could take hours or days, but an equalization charge would speed up the process considerably. The SG in many AGM (absorption glass mat) batteries may be as high as 1.365 but there is no practical way to measure it. If you get new batteries, you should fully charge them, equalize them, and **then** take a specific gravity reading for future reference, as not all manufacturers use exactly the same SG, and SG may also vary for the same battery sold in different climates.

A hydrometer is an instrument used to check the specific gravity of the electrolyte in the battery. Most lead-acid batteries will be in the range of 1.1 to 1.3 kg/l specific gravity, with most fully charged batteries being about

1.23 to 1.30 kg/l. Hydrometers are inexpensive and can be purchased at any car accessory store. Some batteries manufactured for use in very hot or very cold climates may have stronger or weaker acid. If so, it is usually marked on the battery. The energy efficiency of a battery is a function of the discharge current (see Figure 4.2). High discharge currents cause lower efficiencies, shorten the lifetime of a battery, and should be avoided.

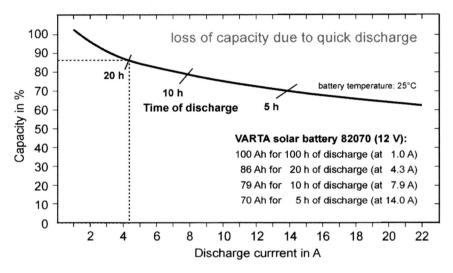

Fig. 4.2. Useable capacity as a function of discharge current for a 12 V lead acid solar battery with a rated capacity of 100 Ah (for 100 hrs of discharge).

4.1.4 Operating Temperature

The lifetimes of lead acid batteries are reduced by 50% for every 10 K above 25°C. Due to the large heat capacity of the battery, peak operating temperatures of the battery are lower than maximum ambient temperatures. Temperature compensation for the charging regime is required in applications where the temperature varies constantly more than ±5 K. It is recommended that the temperature sensor for the operating temperature of the battery is mounted at the positive pole of a battery and is thermally insulated against ambient temperature.

4.1.5 Self-Discharge

Self discharge occurs when a battery is in an open circuit - principally due to interaction between the electrodes and the electrolyte.
Lead from the negative electrode transforms into lead sulfide:

$$Pb + H_2SO_4 \Rightarrow PbSO_4 + H_2$$

At the positive electrode lead dioxide is reduced to lead sulfide also:

$$2\ PbO_2 + 2\ H_2SO_4 \Rightarrow 2\ PbSO_4 + 2\ H_2O + O_2$$

additionally a corrosion reaction with the grid of the positive electrode occurs and builds up a sulfatation layer:

$$Pb + PbO_2 + 2\ H_2SO_4 \Rightarrow 2\ PbSO_4 + 2\ H_2O$$

This self-discharge reaction leads to gas evolution and to a reduction in the capacity of the sulphuric acid concentration. The rate of self-discharge becomes greater as temperatures rises (see Figure 4.3) and the number of charging/discharging cycles increases. This is particularly true with batteries with grids containing antimony.

The float voltage is the tension at which the battery is "floated" or when just enough current is supplied to equal the self-discharge of the battery. This is typically about 14.2 V to 14.5 V for a 12 V battery.

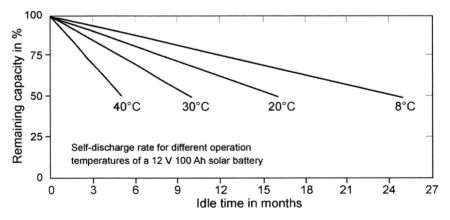

Fig. 4.3. Relative self-discharge of a 12 V 100 Ah lead-acid solar battery as a function of idle time and operation temperature.

4.1.6 Deep Discharge

While other types of batteries (e.g., NiCd batteries) can be used to their full capacity, this should be avoided with lead acid batteries. The greater the "depth of discharge" (DOD) the shorter the lifetime of the battery (see Fig. 4.4).

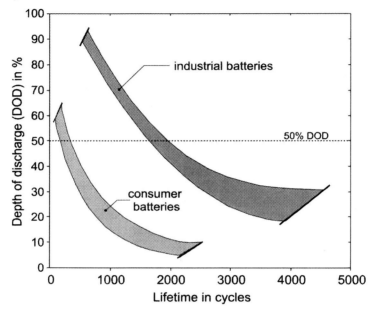

Fig. 4.4. Decrease of useable lifetime for different types of lead-acid batteries as a function of increasing depth of discharge (DOD).

This is due to the reduced sulfuric acid concentration with increasing DOD. In dilute H_2SO_4, solubility of $PbSO_4$ (the primary discharge product) is very high. As a result of sulfating of the electrode, short circuiting and strong corrosion are observed.

To reduce these problems the battery is permitted to discharge down to 50% DOD with batteries used for starting, lighting and ignition (SLI- or automotive-batteries) and 80% DOD using industrial batteries. However, it is very difficult to determine this final DOD. Most current solar controllers act by switching off the charge current at a fixed discharge voltage (cutoff voltage) which is approximately equivalent to the critical DOD. However, the cutoff voltage is load-dependent, only a few solar *I-I* chargers can take account of this, such as the "Solarix" solar controller produced by *Steca*, Germany.

In large PV installations a diesel generator is switched on if the critical DOD is reached (hybrid PV-Diesel system). For smaller battery units a back-up system does not make sense, although over-sizing the battery can be a practical way to avoid deep discharge. Large solar lead acid batteries are rarely supplied and are very expensive. Some new solar batteries are now being developed for this purpose, such as one made by *Akku Gesellschaft*, Germany, with a rated capacity of 240 Ah at 12 V at and concentration of 1.22 kg/l where the capacity increases to 260 Ah if the sulfuric acid concentration rises to 1.28 kg/l. This battery has a life expectancy from five to six years or 800 cycles (50% DOD).

4.1.7 Sulfation

Sulfation is the formation of large $PbSO_4$ crystals, the discharge product of the lead-acid battery. Even though lead sulfate is created in the materials of plates during normal discharging, this term is used to describe the generation of a different form (large crystals) of lead sulfate which will not readily convert back to normal material when the battery is charged. It is very difficult to charge both electrodes when these large crystals are present. Sulfation usually takes place after discharge at low currents (i.e., self-discharge currents) as the result of an acid stratification and crystallization. The process also occurs when a battery is stored too long in a discharged condition, if it is never fully charged, or if the electrolyte level has become abnormally low due to excessive water loss from overcharging and/or evaporation. Often sulfation can be corrected by charging very slowly (at low current) at a higher than normal voltage, usually at about 2.4 to 2.5 V per cell at 0.5 to 8 A (depending on battery size). This will gradually remove the sulfation in many cases.

4.1.8 Battery Types

Lead acid batteries are used for a large spectrum of applications: long-term storage, short-term high current output applications, high energy efficiencies, long lifetimes, low weight, low maintenance, low-cost etc., there is no battery which can completely fulfill all the properties. There are several types of products on the market which are described below. While the costs are mainly given by the materials and technology used, they also depend on production capacity, competitors, and popularity of some battery sizes, which could be seen in Figure 4.5.

4.1.8.1 Automotive Batteries

SLI (starting, lighting, ignition)-batteries were created for starting cars. The battery is designed for short time, high power output and not for deep cycling. However, this type is often used for solar applications as it is relatively inexpensive and easily available. The lifetime is relatively short (one to four years). The maximum DOD should not be higher than 50%. If this battery type has to be out of work for longer periods, it should be kept under a trickle charge current or, alternatively, recharged every two months. Seasonal storage is difficult with this kind of battery if no other source of energy other than solar is available. In order to avoid acid stratification, and therefore sulfation, the battery can be charged at the gassing potential for a short time, or charged directly without a solar controller by the PV generator for from 2 to 5 min twice a week. Mechanical shaking of the battery does not help, as the spacing between the plates is too small to ensure remixing.

4.1.8.2 Industrial Batteries

Industrial batteries are used both for traction and stationary applications yet mainly for cyclic applications. This is possible by using thicker flat plates and tubular electrodes. The active mass in the tubular type is armed by a plastic protector in order to avoid mass shedding. The tubular type is frequently used for solar applications, because its specific kWh costs are relatively low.

4.1.8.3 Solar Batteries

These special types are derived from both the SLI battery and the industrial battery by improving their cycle behavior. Many battery companies adopt a modified SLI battery route as the basic technology is relatively inexpensive. The main disadvantages of the normal SLI-battery, namely bad cycling behavior, can be reduced by:

1. using thicker electrodes

2. using electrodes enclosed in special pocket separators thereby avoiding short circuits by shedding of the active mass

3. using an excess of electrolyte, thereby decreasing the drop of acid concentration during discharge

4. using grids made from low-antimony or antimony-free alloys thereby reducing gassing and self discharge.

4.1.8.4 Industrial Battery Variations

When using modified industrial batteries, basic design does not change, but the electrolyte volume increases and the lead content of the grid alloy decreases. In large installations, such batteries are fitted with an automatic water-refilling system and/or an air pump in order to avoid acid stratification.

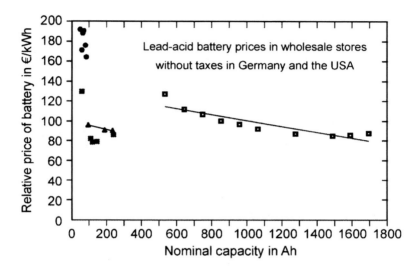

Fig. 4.5. Relative prices of lead-acid batteries as a function of nominal storage capacity (ISET).

4.1.8.5 VRLA ("Maintenance-Free") Batteries

When overcharged, batteries produce hydrogen and oxygen and there is also a consequential loss of water. In open batteries that loss needs to be made up from time to time by refilling the battery with distilled water. Sealed units, when properly operated, minimize this loss; for this reason these are generally considered as "maintenance-free" batteries or VRLA (Valve Regulated Lead-Acid) batteries. If they are overcharged, a valve will let the battery vent, which results in permanent loss of capacity since water cannot be added to these units.

To reduce this effect recovering can be attempted: The oxygen developed at the positive electrode is transported to the negative electrode, where it is reduced back to water. The transport of the oxygen through the sulfuric acid takes place through gas pores formed in silica or glass-fiber used to immobilize the acid. Because most of the electrolyte is immobilized, the acid stratification and the sulphation are reduced. VRLA-batteries were not developed particularly for solar applications but can be used for that purpose. However, a well-maintained open battery has a longer lifetime than a VRLA-battery.

4.2 Other Type of Batteries

4.2.1 Nickel Cadmium Battery

Nickel-cadmium batteries are commonly used as rechargeable batteries for household appliances and can be suitable for stand-alone PV systems. They have a number of advantages over lead-acid batteries:

- They have a higher energy density than lead-acid batteries,
- can be fully discharged, eliminating the need of oversizing,
- are more rugged,
- have excellent low temperature performance,
- have low internal resistance permitting high currents,
- can be charged and discharged at a much higher rate (up to 30 minutes)
- maintain a relatively uniform voltage during discharge, and
- have low maintenance requirements;

however they also have some disadvantages:

- they are typically two times more expensive,
- they have lower energy efficiencies (75%)
- If the battery is not used to its DOD and is not frequently totally discharged, it will loose its capacity (so called "memory effect" - in a contradistinction to the lead acid battery).
- Cadmium is a hazardous material, so it is important to make sure that the battery is not disposed but recycled after its useable lifetime.

4.2.2 Nickel Hydride Batteries

These batteries are similar to Nickel Cadmium batteries, but instead of the cadmium electrode they have a metal hydride electrode. This allows some significant advantages:

- They have a 50% higher specific capacity than nickel cadmium cells,
- the electrode material is much less hazardous than cadmium,
- the memory effect is less than for Nickel Cadmium cells.

Formerly these cells had a large self-discharge rate, but improved production methods reduced this problem. Due to the significant advantages this kind of battery may substitute the nickel cadmium battery in the long term.

4.2.3 Lithium-Ion Batteries

These types of batteries are quite expensive, but offer some important advantages:

- high energy density,
- high efficiency (energy efficiency 95% for room temperature up to 60°C,
- even at -10°C energy efficiency remains 90% (see Kuhmann et al. 2000),
- low self-discharge rates of 2–4%/month between 25°C and 60°C,
- operate even at very low temperatures.

The sensitivity of Lithium-Ion batteries to under- and over-voltage is mastered by integrated control circuits in the battery cases.

4.2.4 Nickel Iron Batteries

Alkaline-type electric cells are using potassium hydroxide as electrolyte and anodes of steel wool substrates with active iron material and cathodes of nickel plated steel wool substrates with active nickel material. This is the original "Edison Cell."

Advantages:

- Low costs,
- Very long lifetimes (3,000 cycles).

Disadvantages:

- Low energy efficiency (typically 55%),
- Very high rates of self-discharge (typically 40%/month),
- High water consumption,
- Limited temperature range (0°C to 40°C),
- High specific weight/volume,
- High internal resistance.

A high internal resistance causes a large voltage drop across a series connection of battery cells. This also means that the output voltage varies with load and charge much more than other batteries. The power conditioning components have to support this voltage variation. If certain common DC appliances (as a refrigerator) are connected to the battery directly they also have to support these voltage swings without damaging. The high losses in charging and discharging will add an extra 25–40% to the size of the solar panels you will need for the same energy usage.

4.2.5 Comparison

Table 4.1. Comparison of different types of batteries

Type of battery system	Technology	Lifetime for up to 80% DOD in cycles	Electrical Energy efficiency	Self-discharge in %/month	DOD	Storage costs in €/kWh	Energy density in Wh/kg	Energy needs for production per kWh of storage capacity (recycled)
Lead-acid	grid, closed	500	83%	3	40% to 60%	0.221	30	137.1 kWh (29.3 kWh)
	tubular, closed	1500	80%	3		0.174		
	grid, sealed	>600	90%	4		0.226		
	tubular, sealed	1500	90%	4		0.297		
Ni-Cd	closed, pocket	2000	71%	6	60% to 90%	0.333	57 (45–60)	499.2 kWh
	closed, sinter	>2000	71%	10		0.641		
	sealed, pocket	1500	71%	20		0.682		
	sealed, sinter	1500	71%	20		1.030		
	sealed, fiber	3500	71%			0.667		
Ni-MH	Potassium hydroxide as electrolyte				60% to 90%		75 (50–70)	287.1 kWh
Li-Ion	module with integrated control circuits	>500			80%	≥ 1	90 to 115	
Ni-Fe		3000	55%	40		0.338	55	

References:
J. Garche, P. Harnisch: *Batterien in PV-Anlagen*, In: J. Schmid, *Photovoltaik*, 1999.
L. Gaines, *Impacts of EV Battery Production and Recycling*, Technical Women's Symposium, April 29-30, 1996, Argonne, Illinois, USA
L. Gaines and M. Singh, *Energy and Environmental Impacts of Electric Vehicle Battery Production and Recycling*, SAE Paper 951865, presented at the Total Life Cycle Conference, Vienna, Austria, Oct. 16–19, 1995.
T. Brohm, M. Maul, E. Nesissner, *Advanced Lithium-Ion Batteries for Electric Vehicles*, VARTA Batterie AG, Research and Development Center, 65779 Kelkheim, Germany

4.3 Fuel cells

4.3.1 Principle

In principle a fuel cell based on hydrogen-oxygen is simply a reversed electrolysis; if its fuel is supplied (hydrogen and oxygen) it will produce electricity.

A fuel cell consists of two electrodes sandwiched around an electrolyte. Oxygen passes over one electrode and hydrogen over the other, generating electricity, water and heat.

Hydrogen fuel is fed into the anode of the fuel cell. Oxygen (or air) enters the fuel cell through the cathode. Encouraged by a catalyst, the hydrogen atom splits into a proton and an electron, which take different paths to the cathode. The proton passes through the electrolyte. The electrons create a separate current that can be utilized before they return to the cathode, to be reunited with the hydrogen and oxygen in a molecule of water.

A fuel cell system which includes a "fuel reformer" can utilize the hydrogen from any hydrocarbon fuel – from natural gas to methanol, and even gasoline. Since the fuel cell relies on chemistry and not combustion, emissions from this type of a system would still be much smaller than emissions from the cleanest fuel combustion processes.

4.3.2 Types of Fuel Cells

4.3.2.1 Phosphoric Acid

This is the most commercially developed type of fuel cell. It is already being used in such diverse applications as hospitals, nursing homes, hotels, office buildings, schools, utility power plants, and an airport terminal. Phosphoric acid fuel cells generate electricity at more than 40% efficiency – and nearly 85% if the steam this particular fuel cell produces is used for co-generation – compared to 30% for the most efficient internal combustion engine. Operating temperatures are in the range of 204°C. These fuel cells also can be used in larger vehicles, such as buses and locomotives.

4.3.2.2 Proton Exchange Membrane (PEM)

These cells operate at relatively low temperatures (about 93°C), have high power density, can vary their output quickly to meet shifts in power demand,

and are suited for applications, – such as in automobiles – where quick startup is required. According to the U.S. Department of Energy, "they are the primary candidates for light-duty vehicles, for buildings, and potentially for much smaller applications such as replacements for rechargeable batteries in video cameras."

4.3.2.3 Molten Carbonate

Molten carbonate fuel cells promise high fuel-to-electricity efficiencies and the ability to consume coal-based fuels. This cell operates at about 649°C. The first full-scale molten carbonate stacks have been tested, and demonstration units are being readied for testing in California in 1996.

4.3.2.4 Solid Oxide

Another highly promising fuel cell, the solid oxide fuel cell could be used in big, high-power applications including industrial and large-scale central electricity generating stations. Some developers also see solid oxide use in motor vehicles. A 100 kW test is being readied in Europe. Two small, 25 kW units are already on-line in Japan. A solid oxide system usually uses a hard ceramic material instead of a liquid electrolyte, allowing operating temperatures to reach 1,000°C. Power generating efficiencies could reach 60%. One type of solid oxide fuel cell uses an array of meter-long tubes. Other variations include a compressed disc that resembles the top of a soup can.

4.3.2.5 Alkaline

Long used on space missions, these cells can achieve power generating efficiencies of up to 70%. They use alkaline potassium hydroxide as the electrolyte. Until recently they were too costly for commercial applications, but several companies are examining ways to reduce costs and improve operating flexibility.

4.3.2.6 Other Fuel Cells

Direct methanol fuel cells (DMFC) are relatively new members of the fuel cell family. These cells are similar to the PEM cells in that they both use a polymer membrane as the electrolyte. However, in the DMFC, the anode catalyst itself draws the hydrogen from the liquid methanol, eliminating the need for a fuel reformer. Efficiencies of about 40% are expected with this type of fuel cell, which would typically operate at a temperature between 48.9°C and 87.8°C. Higher efficiencies are achieved at higher temperatures.

4.3.2.7 Regenerative Fuel Cells

Still a very young member of the fuel cell family, regenerative fuel cells would be attractive as a closed-loop form of power generation. Water is separated into hydrogen and oxygen by a solar-powered electrolyzer. The hydrogen and oxygen are fed into the fuel cell which generates electricity, heat and water. The water is then re-circulated back to the solar-powered electrolyzer and the process begins again.

5
PV-Systems in the Tropics

While nowadays most PV capacity is installed grid-connected in moderate climates due to a favorable legislation in some countries there, the largest "natural" markets for PV are off-grid applications within the topical "sunbelt" along the equator. The basic layout of a typical PV system for off-grid applications (grid connection plays a minor role in the tropics, most systems are used for rural electrification) is relatively simple: PV generator (consisting of a number of PV modules), energy storage (mostly a lead-acid battery), and power conditioning devices (consisting of a charge regulator and, optionally, an inverter). In total, about one million such systems have been installed worldwide. Nevertheless, in practice frequent failures have been reported; many of them are not of technical nature. Based on ten years of practical experience in Brazil, this chapter highlights some important considerations and best-practice examples of how to implement a reliable, efficient, long-lasting and cost-effective PV power supply. Some of the recommendations given may seem trivial, but apparently have not been considered in many system layouts and installations in the past, even at large scale rural electrification programs.

5.1 Pre-installation Issues

5.1.1 Additional Considerations for Planning

5.1.1.1 Determination of Load Requirements

While the price of a PV installation is increasing roughly linearly with the electricity consumption, customers tend to underestimate design loads during the planning phase in order to save costs, thus making the system sub-dimensioned and unreliable. The best method is to measure power and energy consumption over a period of time. An analysis of consumption frequently shows that it is considerably cheaper to replace older devices with a high energy consumption (e.g., light bulbs, refrigerators, freezers) by new energy-efficient ones than to size the PV generator and the batteries according the energy requirements of the old devices.

5.1.1.2 Dynamics of Project Development vs. Time Constrains

Time from project definition, layout, financing, ordering, shipping to "turn-key" installation can take years (considering bureaucracy for project legislation, import taxation and liberation). Importation often gets extremely time-consuming, and thus costly external services, by so called "despachos," should to be used. Waiting times of 1 to 1.5 years have been reported, so packaging of goods may degrade and equipment may "disappear."

During that whole process side-conditions of the project could have altered: people to whom the power plant was designated could have moved, electrical grid-lines could have been installed, diesel-generating-sets could have been bought; even the planned installation site could be flooded or be transformed into a landfill.

Straightforward planning from the scratch to the turn-key energy supply system rarely happens and often the design requires reevaluation and noteworthy modifications. To accomplish this, constant readiness, observation and controlling are a must.

5.1.2 Financing

Procuring project support programs is an important but time-consuming task. Often a lot of restrictions are bound for such aids: limited time periods for installation, obligatory use of national equipment (e.g., batteries), cooperation with local companies and utilities, long-term power delivery guarantees, certification by local authorities, hard bidding competition. Whereas financial compensation is given in local currency and local interest rates apply, risks associated to that have to be taken into account, such as fluctuation of the exchange rate. A localized component production or assembly-line could reduce these risks, because salaries, rental fees and raw materials are paid in local currency; for most cases this measure will bypass high import taxes.

5.1.3 Importation

Import of equipment and materials is a tricky and time-consuming issue: In order to protect some local industry from cheap imported products import taxes may be high and can have a large impact on the budget (import taxes may double the price of some goods, principally electronics). Knowing the fact that these regulations are not economically sustainable and possibly will paralyze a country's economy on the long term, authorities frequently exclude goods of strategic interest - such as renewable energy technology - from high taxation, but this requires knowledge, extensive paperwork and the declared goods have to precisely fit into the exemption categories.

Recommendations: Make sure that every component of your system is well described (in the local language!). Aside from detailed records of the equipment, the background of the project-partners has to be accurately documented as well (e.g. company's founding certificate, financial and technical capacity proofs). Units mentioned in the documentation (per piece, per volume, per kg, British/American/local measures) are a critical issue, as well as currencies and its exchange rates and which one is to apply (e.g. Brazil has six different official exchange rates for R$ vs. US$). Sometimes parts have to be divided into sub-components: general electronics, processing units, memory, transformers, terminals, instrumentation, displays, PV units, often with different import tax rates and separate forms to be competed and to be certified. Often importation is not taxed upon the value of the imported goods only, but on the total expenditure, including freight costs.

5.1.4 Language Barriers

5.1.4.1 Within Bureaucratic Stages

Fund and credit applications, certification and importation forms require profound knowledge of the local language. Especially juristic and technical expressions may provide difficulties. Terminology could deviate within the "same" language. (e.g. Portuguese from Portugal vs. Brazilian Portuguese or British / American / Indian / African English). Authorities and officials are often quite sensitive in respect of the "correct" use of their terms and in respect of their position. They may deliberately delay processes if they feel offended. On the other hand they don't bother if you frequently ask (in a friendly manner) about the actual project status (a situation where usually European officials would become annoyed). Best strategy is to "involve" the officials in your project, i.e. explain them how important the project is for their country, so they feel proud being a part of it.

5.1.4.2 At the Location of Installation

Aside from the general language barriers, remotely located individuals are often not familiar with basic technical terms. Interactive training is a must, even if it becomes time consuming. The use of well-illustrated training material is important. Illiteracy is widespread (often more than government statistics indicate), especially at remote sites where children are needed for work. You cannot expect to meet people that speak a foreign language on-site.

5.2 Technical Issues

5.2.1 Mounting

Harsh environmental conditions such as high temperatures, high humidity, high irradiance, strong winds (at locations close to the sea connected with salty air), and difficult accessibility for repair requires the use of best avialable materials on-hand. Often stainless steel components are not available locally and have to be imported.

5.2.1.1 Fixation of PV Modules

Mounting and fixing should preferably be carried through the utilization of rivets instead of screws in order to avoid becoming loose and deter theft. Additionally, stainless steel screws may be difficult to obtain, and may even be subject to theft (see below).

5.2.1.2 Wiring of PV Generator

Local electricians do not generally work with high current, low voltage DC wiring (besides car electricity) and therefore flexible UV-resistant cables with an adequate diameter and color are difficult to obtain. The same is true for switches, fuses and terminals. Often these parts are not available locally and must be imported.

5.2.1.3 Theft Prevention

Increasing theft of PV panels is a serious issue in tropical locations, especially in Africa. South African Telecom decided to acquire exclusively orange PV modules in order to ease identification of stolen PV panels. Components may be mounted by rivets, rather than screws in order to make theft more difficult (see above). Often even smaller components, such as instrumentation, terminals, plugs, cables or even water tabs are subject to theft, because for poor people these objects already represent a huge value. A community-adaptation of the system and the nomination of a responsible local person in-charge are most helpful to avoid this issue.

5.2.1.4 Safety Considerations

Aside from the save installation with isolation of terminals and wiring, the batteries should be stored in a secure location, apart from the living area. Metallic objects deposited on the battery terminals may provoke severe accidents.

5.2.2 Non-MPP Operation of PV Generator

Due to the typical *I-V* characteristics of solar cells and PV modules, electrical power output decreases in an almost nearly linear manner when the load curve meets the *I-V* curve of the PV generator on the "left" side of the Maximum-Power-Point (MPP), towards lower operating voltages. On the other side, power declines exponentially when the load curve cuts the *I-V*-generating curve on the "right" side of the MPP, towards higher operating voltages. The latter occurs when the voltage of the generator is reduced (e.g. for high operational temperatures at single- or multi-crystalline silicon solar cells, which typically feature a voltage-temperature coefficient of -0.4%/K to -0.5%/K).

Therefore, it is recommended to either acquire PV modules that contain a larger number of solar cells in series connection in order to increase the generator's voltage level, or obtain charge controllers with an integrated step-up DC-DC converter and a MPP-tracker in order to enable full battery charge even when a generator's voltage levels are reduced. Unfortunately the industry does not supply such equipment as a standard product.

5.2.3 Energy Storage

5.2.3.1 Battery Types

while there are many different types of batteries available, lead-acid batteries nevertheless offer the best cost-benefit ratio. Lead-acid batteries are offered in three versions: conventional ("open" – allowing replacement of lost water), "maintenance free" (surplus of electrolyte, reduced gassing), and "sealed gel" (absorbed electrolyte). Whereas maintenance practice is often poor or the replacement of water is carried out with contaminated water, maintenance-free batteries are preferred. Fully sealed batteries are less robust and have shown shortened lifetime in comparison to maintenance-free batteries in warm climates, especially when charge-control is not properly carried out.

5.2.3.2 Nominal Voltage Level

In most tropical countries solar batteries are derived from car or truck batteries and thus available only in 12 V blocks. The maximum storage capacity to be found in Brazil is 220 Ah and thus the batteries have to be paralleled in order to achieve a greater storage capacity at a given voltage. Due to tolerances in the production process of the batteries (e.g. at the electrode plate treatment or within a battery's acid concentration), differing electrical characteristics such as internal resistance are likely to occur. In this case, paralleled batteries will suffer from considerable internal currents that result in reduced storage capacity and shortened lifetime of the battery bank. We experienced considerable voltage differences even at more expensive, high capacity batteries. That effect may be caused by a poor fabrication process and the lack of quality control, possibly due to low production number of high capacity batteries. Usually test procedures are difficult to be carried on the installation-site. Batteries also may develop problems due to non-homogeneous storage conditions: exposure to high temperatures (e.g. due to exposure to sunlight) cause additional internal currents and damage.

Because import taxes for batteries are often prohibitively high, for most cases the only solution is to use national products with higher voltage levels – thus eliminating the need for paralleling of batteries. Using a 12 V, 220 Ah battery at a 50% maximum depth of discharge, the voltage level has to be increased by an additional 12 V for each 1.3 kWh of storage requirements. For voltages above 50 V additional safety concerns must be addressed. Conventional lead-acid batteries have to be exchanged about four times during the lifetime of a PV system. To keep the system working, a replacement strategy has to be established to have the funds available for the purchase of a new battery and the manpower to replace it. Some refurbishing companies are willing to pay a small amount for old batteries, but unfortunately these companies are located far away from most projects in the big cities and are specialized on automotive batteries.

5.2.4 Power Conditioning Equipment

While in developed countries most electrical machinery used as a load features a cos ρ between 0.9 and 1, local equipment may show a cos f as low as 0.6, so the inverters have to be capable to handle such inductive loads. Also friction losses tend to be higher that increase starting currents for the inverter.

5.2.4.1 Switching Devices

Many inverters are driven by MOSFETs. Whereas on one hand MOSFETs can be easily paralleled due the positive temperature coefficient of their on-resistance, this could lead into a vicious circle when the device operates at elevated temperatures: elevated temperatures increase the internal on-resistance and lead to additional more heat-dissipation that will heat up other paralleled MOSFETs mounted on the same heat-dissipating device.

5.2.4.2 Ventilation

Many inverters use forced ventilation via temperature-controlled fans. Often on-demand fans fail to initiate; jammed by leaves or remains of insects. To avoid that, "cool" operation is recommended, using over-dimensioned switching devices (featuring lofty current reserves and low on-resistances) that require just natural ventilation. It also has to be considered that surrounding air-temperature could easily reach 60°C in a container, so thermal layout has to be made for operation temperatures of 70°C and above. Usually maximum operation temperature for power conditioning equipment is 55°C, resulting in shortened lifetime and failures (e.g. capacitor de-lamination, display degradation, terminal melting and eventually short-circuiting).

5.2.4.3 Charge-Controllers

Non-temperature controlled charge controllers may cause excessive gassing at high operating temperatures. This leads to a loss of electrolyte which should be replaced. However, this is not possible at "maintenance-free" and sealed batteries, so capacity and lifetime will be reduced.

5.3 Operation and Maintenance

5.3.1 Pollution and Degradation of System Components

While the fauna & flora is prevalently richer in topical locations, it will be more likely that it will affect components of a remotely located PV system: bird droppings, seeds, pollen, leaves, branches, dust and dirt spots may accumulate on the PV panel and lead to significant performance losses. Usually the solar cells inside a PV module are connected in series, so the most shadowed cell dominates the output current of the circuit. If bypass diodes are not properly applied, permanent damage of the cells by "hot spots" may occur. The self-cleaning effect requires an adequate inclination angle of the panel, but at tropical locations the PV-panel is mounted almost horizontally to capture a maximum of irradiance, thus inhibiting self-cleaning to a large extend. In the shelter for batteries and power conditioning equipment birds, mice, rats, snakes, spiders, cockroaches, termites, anions, scorpions, frogs, lizards and bats may find their home also may corrupt cables, terminals and relays within a short time period (Fig. 1.5). Also the psychological effect has to be taken into account: if the location is dirty and smells, maintenance tasks are not attractive and will not be carried out.

Recommendations: On one side the hut has to offer good ventilation to keep temperatures at a supportable level and allow the escaping of battery gases, on the other hand the invasion of insects and animals has to be prevented (e.g. grids, solid nets). Keep it nice and tidy.

5.3.2 Monitoring

Local personal may have difficulties identifying and describing the state of the system (or are intimidated to do so) and are hardly capable to to carry out the appropriate actions. Thus, training and capacity building of the local personnel is essential. Additionally, adequate easy-to-read instrumentation and logical, easy-to-explain, fail-save switches with optical and/or acoustical feedback – best within a flow diagram of the system – as well as are the long-term supply of spare parts and adequate tools. A local reponsible person in charge of the plant has to be nominated and paid; this measure also prevents vandalism and theft (see above). A remote monitoring system (via cellular phone line or via satellite) could be tremendously helpful to identify problems or to initiate preventative maintenance.

5.3.3 Further Recommendations

Each successful implementation counts much more worth than a lot of advertising and many words. Considering the recommendations given, chances are pretty good to carry out an effective project; in order to keep it operating, keep the local person in charge motivated, by supplying him/her with sufficient documentation, spare parts, an adequate salary and maintain communication with him/her. Frequent visits secure that the system stays in good shape. Even if the cost/befit ratio of your project is decreasing – keep it working: consider it as a demonstration project and set a positive example. Potential new clients, in form of visitors of the plant, are very likely appear. New projects may come into sight just by carefully observing the environment. Often solutions seem to be obvious, but are not carried out (see Fig. 5.1).

Fig. 5.1. Diesel powered generator for the supply of a 2 kW cellular phone repeating station with noise and exhaust gases in a hotel area in Cancun, Mexico. Operation personnel has to arrive every 12 hours to do a refill – why not PV?

5.4 Concluding Remarks for PV in the Tropics

While conditions are principally very favorable for the implementation of PV at tropical locations, such as high irradiance with low seasonal variation and the necessity for rural electrification, many obstacles have to be overcome. Considering the recommendations given above, the chances are good to implement a successful project. Whereas not each and every aspect could have been analyzed into its full detail, moreover abilities for improvisation, persistence, compassion and good humor are essential to carry out PV projects in the tropics.

6
Energy Consumption for the Set-Up of a PV Power Plant

The "cumulated energy expense" (*CEE*) of an energy supply system consists of the material and energy fluxes needed during manufacturing (CEE_M), operation (CEE_O) and recycling (CEC_R). This chapter deals with the energy expense to manufacture a PV power plant (see Fig. 6.1) including mining, processing, and production of materials for operation and supply, including transportation.

6.1 Preliminary Remarks

Often, the literature is based on data recorded when the production facilities of the manufacturers have not been entirely used (50-70%) for production, e.g., due to market saturation from 1995 to 1999. While energy consumption for many processes is determined by a massive base load and a relatively small amount of processes are dependent directly on the quantity of fabrication, the observed specific energy consumption is considerably higher than for full capacity conditions (see also Hagedorn 1989).

To take account of these parameters, three different case models have been considered. The individual components of cumulated energy consumption for PV power plants CEE_M are indicated by the variable W for clarity and to avoid multiple indices.

6.1.1 Differentiation of the Model Cases

The first case model refers to the existing production technology standards, operating one shift (8 hrs) at full utilization, with an output for cells and modules of 2 MW_p/a. The second case model examines the technological improvements in production together with an expansion to a four-shift operation, resulting in a production capacity of 25 MW_p/a.

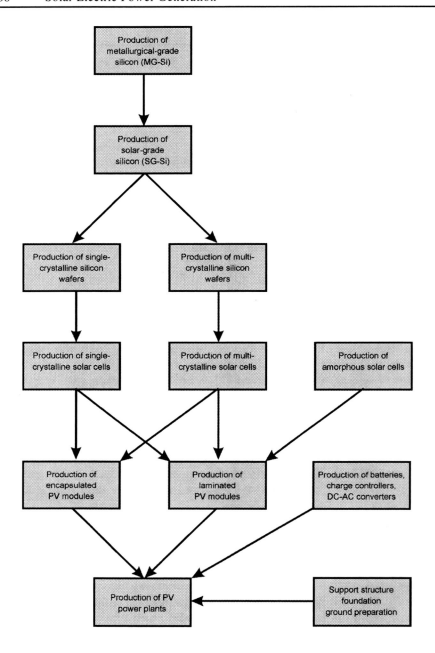

Fig. 6.1. Significant steps for the production of PV power plants (without depletion of raw material) relevant for the manufacturing energy expense (MME) using different production technologies.

6.1.2 Equivalent Primary Energy Consumption

To calculate the accumulated primary energy consumption, the values of energy consumption for production have to be converted into the equivalent primary energy values. For electrical energy a degree of utilization (primary energy efficiency) of 0.4 was used, for gaseous and liquid fuels that value was 0.85 and for non-energetic consumption 0.8. These conversion factors are taken from the structure of energy supply in Germany and represent the conservative (lower) limit of the known values. Additionally, the effect of displacement of locations for production and for application of PV power plants have been examined for the first time. These results could also be applied for the production of internal energy generation such as solar thermal systems, wind generators, PV or diesel generator with heat recovery.

6.2 Preparation of Raw Materials for Production

The raw materials necessary for production require a certain expenditure for opening up and development of the ore deposit (W_{res}), as well as directly for release and exploitation (W_{free}). Materials must be transported to the location of processing (W_{transp}). Further processing to meet operational standards for the materials require additional energy (W_{prep}). The sum of these is the participation of energy consumption and emission before the proper production of the actual PV power system ($W_{pre\,pro}$).

$$W_{pre\,pro} = \sum W_{res} + \sum W_{free} + \sum W_{trans} + \sum W_{prep} \qquad (7)$$

6.2.1 Development of a Deposit

Depending on the kind of energy carrier or raw material and the local conditions, the expenses for the development of the deposit could be considerable. For example, oil exploitation, both prospecting and probe drills are necessary ($W_{prospect}$). Often, only every fiftieth drilling is successful. Also transport of the equipment (W_{trans}) is required (tower, tubes, tools), the set-up of an infrastructure ($W_{infrastr}$) at the deposit (roads, rail, pipelines, housing, airports) has to be taken into account. Underground mining requires shaft fitting (W_{prep}). For open pit mining dislocation of villages is sometimes necessary. Last, but not least, after exhaustion of the deposit, the site has to be cleaned and restored to its original condition ($W_{recycle}$).

$$W_{res} = \sum W_{prospect} + \sum W_{infrastr} + \sum W_{trans} + \sum W_{prep} + \sum W_{recycle} \qquad (8)$$

6.2.2 Release (Exploitation)

The raw materials have to be released from the ores and separated ($W_{separate}$). The technology used is dependent on material density and occurrence. For solid raw materials that process could be carried out mechanically (by hammers or excavators) or chemically (as coal liquefaction). Liquid or gaseous raw materials can be removed from the ores by pumps or escape themselves. Afterwards, the raw materials must be transported by conveyer belts, elevators, tubes or pipelines to the Earth's surface. Often, the materials are impure and only a fraction of the materials released can be used (W_{sort}).

$$W_{free} = \sum W_{seperate} + \sum W_{trans} + \sum W_{sort} \qquad (9)$$

For the example of iron ore exploitation the most important appliances and wear-parts (in the sequence of process-course) are quoted in Table 6.1. The specific operation weight of an equipment is given by its weight related to its throughput during its useable lifetime.

Table 6.1. Material expenditures for equipment for ore exploitation (Schaefer 1993)

Device	Operation weight in metric tons (t)	Specific operation weight losses (depreciation of materials) in g t_{steel} / t_{ore}
Rope excavator	3,300	116
Conveyer belt (7 km)	181	12
Paddle wheel excavator (for open bit mining)	14,300	8
Round crusher	119	2
Impact crusher	37	4
Wear parts	< 1	429

Table 6.2. Required earth- and ore movements for different materials (Mauch 1995)

Material	Weight of ore required in relation to weight of final product in $t_{ore} / t_{product}$ (metric tons)
Copper	1000
Aluminum	13.7
Raw iron	4.4
Cement	1.7

6.2.3 Transport

As well as the direct costs for taking the ore to the surface, the transport to the processing plant and to the consumer has also to be considered. Tables 6.3 and A8 (in the Annex) show the different means of transport with their specific energy needs and carbon dioxide emissions. The actual length of transport has to be accounted for, while not always the shortest route is taken due to technical, economic and political reasons (e.g., Suez canal and Panama canal).

Table 6.3. Energy consumption and CO_2 emissions for transport of freight

Type of transport	Direct energy consum. in MJ/(t km)	Direct CO_2 emissions in kg/(t km)	Direct plus "grey" energy cons. in MJ/(t km)	Total CO_2 emissions in kg/(t km)	Type of fuel	References
Freight ship (overseas)	0.09	0.0075	0.12	0.0094	100% HO	Frischknecht 1994
Container ship 47,000 t 43 km/h	0.029	0.0069	0.12	0.0086	100% HO	Frischknecht 1996
Freight ship (inland)	0.48	0.04	0.84	0.063	100% D	Frischknecht 1994
	0.51	0.0365	0.844	0.05863	100% D	Frischknecht 1996
Water (general)			0.4	0.03		Lenzen 1999

92   Solar Electric Power Generation

Type of transport	Direct energy consum. in MJ/(t km)	Direct CO_2 emissions in kg/(t km)	Direct plus "grey" energy cons. in MJ/(t km)	Total CO_2 emissions in kg/(t km)	Type of fuel	References
Rail [1]	48	0.0556	01.39	0.1137	51% E 49% D	Frischknecht 1994
	319	0.0104	0.956	0.0513	80% E 20% D	Frischknecht 1996
Rail (gen.)			0.9	0.08		Lenzen 1999
Road (40 t) 50% loaded	116	0.967	2.16	0.1606	100% D	Frischknecht 1994
	1	0.079	2.08	0.13637	100% D	Frischknecht 1996
Road (gen.)			1.7	0.13		Lenzen 1999
Air			65.10	5.40		Lenzen 1999

Direct energy consumption is given by fuel consumption; indirect or "grey" energy consumption also considers wear out of means of transport and infrastructure (roads, rails)
Detailed data are given in Table A8 in the Annex.

[1] includes shunting and shifting
CO_2 emissions of Frischknecht 1994 have been calculated using the following CO_2 intensities:
electricity: 0.53 kg CO_2/kWh (Germany), fuels: 0.3 kg CO_2/kWh, non-energetic consumption (NEC, also "grey" energy): 0.23 kg CO_2/kWh (see Wagner 1996)

G: Gasoline   HO: Heavy oil
D: Diesel fuel   E: Electrically driven

6.2.4 Preparation for Production

Very rarely raw materials can be used directly; in most cases separation (W_{sort}) and processing ($W_{process}$) is necessary. The resulting waste materials have to be disposed or recycled (W_{waste}).

$$W_{prep} = \sum W_{sort} + \sum W_{process} + \sum W_{waste} \qquad (10)$$

6.2.5 Construction Work and Buildings

The equipments for production of the raw materials and the manufacturing plant itself require construction measures, e.g., buildings. The set-up of these has to be included in the balance of the production of the PV power plant. The useable lifetime of a building is difficult to estimate: modernization, closing-down or war occurs before the useable lifetime of the building is over (about a 100 years). The set-up costs of walls are given in Table 6.4, that of windows in Table 6.5. In the case of substitution of building structures by photovoltaic elements (e.g., by PV facades, as shown later) these tables are helpful when considering construction costs that can be avoided in the balance.

Table 6.4. Energy requirements for the production of external walls (Geiger 1993)

Composition of wall	Wall thickness in cm	Insulation thickness in cm	Weight in kg/m²	Thermal conductivity k in W/m²K	CEE_M in MJ/m²
Single brick wall (light high hole bricks)	40	0	440	0.76	972
Thermal insulating compound wall [1]	34	7	395	0.38	1220
Thermal insulating compound wall [1]	37	10	395	0.29	1277
Thermal insulating compound wall [1]	42	15	396	0.20	1372
Two shell wall [2]	49	7	614	0.34	1742
Two shell wall [2]	57	15	615	0.19	1894

[1] bricks with vertical cavities holes and thermal insulation layer of styrofoam® (expanded polystyrene)
[2] bricks with vertical long cavities and font wall bricks with a thermal insulation layer of styrofoam® (expanded polystyrene)
CEE: Cumulated Energy Expenses

Table 6.5. Energy requirements for the production of windows (Geiger 1993)

Material for frame	Energy requirements per m² of window area in MJ/m²	
	2-sheet insulating glazing $k = 2.6$ W/(m² K)	heat preservation glazing $k = 1.4$ W/(m² K)
Wood	934	1088
Plastic	1331	1485
Aluminum	3187	3340

The expenses for the ground moving applications necessary for the set-up of the building are shown in Table 6.14, page 114. For cement Frischknecht (1996) gives a CO_2 emission rate of 0.88 kg per kg cement; further detailed data of energy requirements and CO_2 emissions for concrete (consisting 12% of cement, 82% of sand and gravel plus 6% of water, by weight) or for other construction materials can be found in Table A5 in the Annex, pages 237-239.

The remarkable fact is that the accumulated operational energy requirement after 50 years of use for common, relatively new buildings (built according to the German Heat Preserving Edict of 1982) is 75% of the total primary energy requirement. For buildings built considering the latest heat preserving edict of 1995 that operation energy requirement will only be 60%. The total primary energy requirement of such new buildings could account for 40 GJ/m² (Geiger 1993). In the future, energy saving during the construction phase could play a more important role than the current discussion which is focused on the reduction of operational costs – which is nevertheless very important.

6.3   Direct Energy Consumption at the Production Process

To consider the energy demand for the production process of PV power stations, the energy requirements for the solar cells (W_{cell}), the solar module (W_{module}) and electrical power conditioning units (W_{conv}) have to be taken into account as well as the demands for the support structure and the foundation ($W_{support}$).

$$W_{pro} = \sum W_{cell} + \sum W_{module} + \sum W_{conv} + \sum W_{support} \qquad (11)$$

6.4 Production of Solar Cells

Because most terrestrial PV applications are based on silicon solar cells, the following considerations are focused on them. Silicon is the second most abundant element of the Earth's crust and therefore almost unlimited. Silicon has a stable crystalline structure, so cells have a very high lifetime (> 25 years). The band gap of silicon is relatively well suited to convert the Sun's spectrum to electrical energy. The disposal of silicon is similar to that of glass and, thus, without serious problems. The main challenge for production is the purifying process which should result in quite pure silicon to allow efficient solar cells, but should require as little energy as possible.

6.4.1 Production of Technical Silicon (MG-Si)

Technical silicon or metallurgical silicon (MG-Si) is produced in large scale for the demands of the aluminum and steel industry. An order of one million metric tons is produced globally every year. The raw material silicon oxide (SiO_2) in form of quartz or sand is reduced by carbon to silicon, while emitting CO or CO_2. The reduction is carried out in large arc furnaces by carbon (in the form of a mixture of wood chips, coke and coal) to produce silicon. The liquid silicon at a temperature of 1,500°C is periodically poured from the furnace (at purity of 98–99%) and blown with oxygen or oxygen/chlorine mixtures to further purify it (up to 99.5%). It is then poured into shallow troughs, where is solidifies and is subsequently broken into chunks. Production capacity of a typical furnace is one ton of MG-Si per hour. Using a novel way of feeding (mixing of pieces of quartz and sand briquettes) the electrical power consumption could be reduced to 13 kWh/kg. Transport, production of feed mixture, feeding, and grinding are considered by 1.6 kWh/kg.

Combustibles at the level of energy consumption are mainly coke and coal which are used for the sand briquettes. Energy losses of more than 50% are caused by emissions. This offers a great opportunity for reductions, but also bears a great technical and economical risk, and is therefore not yet considered.

6.4.2 Metallurgical-Grade Silicon (MG-Si) to Semiconductor-Grade Poly-Silicon (EG-Si)

For the use in solar cells (as well as for other semiconductor devices) silicon has to be much purer than metallurgical grade silicon. The standard approach for purification is known as the "Siemens C-process." After grinding (grain size < 0.5 mm) the metallurgical silicon is fluidized in a reactor at 300°C to 400°C with HCl in the presence of a Cu catalyst generating $SiHCl_3$ and H_2. The gases are passed through a condenser. The resulting liquid is subjected to multiple fractional distillation to produce *tri*chlorosilane $(SiHCl)_3$. To extract the pure silicon the $SiHCl_3$ is reduced by hydrogen in a reactor where the silicon deposits itself at electrically heated silicon rods of 1,000°C in a fine-grained polycrystalline form. The latter step not only requires a lot of energy, but also formerly had a low yield (ca. 37%, according to Green 1986).

Production of semiconductor-grade silicon is done in large scale (ca. 3,000 metric tons/a) in compound chemical production plants and therefore has to be considered in energetical and material association with the whole product range of the plant. For example, the use of waste *tetra*chlorsilan $(SiCl_4)$, which reduces the amount of *tri*chlorsilane $(SiHCl_3)$, has to considered. The heat from the reactors during the cooling stages is fed into the vapor pipe grid of the production plant and used for other products. This leads to substitution of fuels and therefore creates an energy bonus of 29.8 kWh/kg for combustibles. The electrical consumption of the process is 114.3 kWh/kg. While the process still offers a potential for optimization from the technical as well as from the energetic point of view, significant improvements cannot be expected in the present economical conditions.

6.4.3 Production of Single-Crystalline Silicon

For the semiconductor electronics industry, silicon must not only be very pure, but it must also be in single-crystal form with zero defects in the crystal structure. The major method used to produce such material commercially is the so called "*Czochralski*" process. The semiconductor-grade silicon is melted in a crucible with trace levels of a dopant. For solar cells, boron, a p-type dopant, is normally used. Using a seed crystal in a silicon melt a large cylindrical single crystal can be grown by slowly pulling it from the melt. The resulting single-crystalline rods have a typical diameter of 125 to 130 mm and a length of 1.5 to 2 m. The pulling is carried out at a speed of 1 mm/min, so the melt has to be maintained at 1,450°C for many hours. Even with very good thermal isolation high energy losses are reported due to heat transfer by thermal radiation of the surface of the melt. An increase of the yield could be only carried out by an increase of the diameter of the rod, which can be done in an industrial manner to diameters up to 212 mm (eight inches), see also Aulich 1986. The round rods are very often sawn to columns with a quasi-quadratic cross section (quadrats with round corners), see Figure 6.3, to increase the fill of the PV module area.

Beside the *Czochralski* process, the *floating zone* (FZ) process exists, which allows for the production of highly efficient solar cells: Polycrystalline ingots are directed though an inductive heating ring which melts the column locally. While the solubility of most impurities in the melt is better than in the solid body, they diffuse into the molten zone. By moving the molten zone along the whole column towards the end of the crystal, the impurities can be taken off from the middle part and be accumulated at the ends (which are sliced off later). Heating and crystallization is carried out without touching the silicon. Repeating this process several times produces the purest silicon. This process requires a lot of time and energy which makes FZ-Si based cells extremely expensive and thus their use is mainly limited to space applications.

6.4.4 Semiconductor-Grade Silicon to Multi-Crystalline Silicon

EG-Si is also molten under a low pressure, protective atmosphere, but is then simply poured into a crucible. The cooling and crystallization process is temperature controlled. This cast silicon does not have the same high quality as the material produces by CZ or FZ, but is good enough to make quite efficient solar cells [17]. Rather than being a large single crystal, the ingot is made up of many smaller crystals, or "grains". A typical crystallization rate is 0.5 kg/min. The ingots are cut to columns with a square base, and are sliced later (see Figure 6.2). According to Wagemann 1994, the conventional Silan production method for multi-crystalline wafers requires has a specific energy requirement of 2,300 kWh/kg (at a yield of 8 %, incl. slicing by conventional inhole saws, see also page 98. Using lye bath processing with solar grade silicon (at a yield of 35%), cucible casting and multi-wire slurry slicing only 200 kWh/kg are required.

Table 6.6 gives an overview of the typical energy requirements for the silicon growth and possible efficiencies for solar cells based on that material.

---

[17] Industrial multi-crystalline silicon cells of 100 cm² area have reached 15.8% conversion efficiency (Sharp). The laboratory record is 17.8% for a 4 cm² cell (UNSW). Commercial multi-crystalline cells are about 15% efficient (e.g., Q-cells).

Table 6.6. Energy requirements for different Silicon growth methods (Schmela 2000)

Method	Width in mm	Growth rate in mm/min	Ingot through put in m²/day	Energy use[1] in kWh/kg	Energy use[1] in kWh/m²	Best effi- ciency[2]	Typical efficiency[2]
Floating Zone (FZ)	200	2–3	80	30	36	24%	<18%
Czochralski (CZ)	150	0.6–1.2	30	18–40	21–48	22%	<15%
Directional Solidificat. (DS)	660	0.1–0.6	70	8–15 [3]	9–17	18%	<14%
Electromag. Casting (EMC)	350	1.5–2	600	12	35	16%	<14%
Dendritic Web	50	12–20	1	-	200	17%	<15%
Capillary Die Growth (EFG)	800	15–20	20	-	20	16%	14%
Edge-Supported Pulling	80	12–20	1.7	-	55	16%	<13%
Substrate Melt Shaping	200	1000–6000	>1000	-	-	16%	<12%

Remarks [1] energy use for crystal growth
[2] for solar cells based on that material; if SiNx:H is used as anti reflective coating, cell efficiencies can be 1.5% higher.
[3] typical: 12 kWh/kg (given by major manufacturer GT Solar, 2006)

6.4.5 Production of Silicon Wafers (Single- and Multi-Crystalline)

For photovoltaic purposes, silicon solar cells need to be only 0.1 mm or so thick to absorb most of the appropriate wavelengths the sun's radiation. Therefore the large single crystal column (*ingot*) can be cut up into slices (*wafers*) of the same thickness. The often-used in-hole saws result in cutting-losses of 50% and at minimal wafer thicknesses of 0.45 mm (independently whether the wafers are single- or multi-crystalline, see Figures 6.3 and 6.2).

The use of slurry-wire-saw technology allows the production of thinner wafers (0.2 mm) with a considerably superior cut (cutting-losses of 0.1 mm only) and less crystalline distortions. This reduces the depletion of silicon and the resultant energy consumption. Furthermore several hundred wafers can be cut at the same time – even the whole ingot (up to a length of 1.6 m) can be sliced into wafers. In-hole saws produce only one wafer at a time. At the slurry-wire-saw, a wire of 0.18 mm of thickness and of up to 200 km of length is stretched by four guide rollers to a square while it gets stripped from a spool with a velocity of 5 m/s. The machining is done using a silicon carbide with a grain size of 12 μm, suspended in cutting oil. The oil serves as a carrier of the silicon-carbine as well as for heat transfer. The thinner silicon wafers suffer from high brittleness and have to be treated with great care.

Fig. 6.2. Ingot of multi-crystalline silicon (formerly known under the name of "polycrystalline silicon") with square cross-section.

Fig. 6.3. Ingot of single-crystalline silicon with quasi-square cross-section and a sliced wafer in the foreground (Photos taken at HMI, Berlin).

6.4.6 Single-Crystal Wafers to Single-Crystalline Solar Cells

On the way to becoming a working solar cell the single-crystalline wafers undergo the following process:

- **Cleaning** via "lap" (fine polishing with Al_2O_3 with a grain size of 10μm as abrasive media), to remove damage caused by the wafer slicing process.

- **Texturization** via damage-etching (wet-chemical with KOH as cauterant) to generate a pyramid-like structure serving as optical anti-reflective layer (or by sputtering of TiO_2 or Ta_2O_5 in vacuum and sintered at 400°C – after deposition of the contacts),

- **Doping:** It was mentioned in the preceding section that, in standard solar cell technology, boron was normally added to the Si-melt, which then resulted in p-type wafers. To form a solar cell, n-type impurities are introduced to generate a p-n junction. Phosphorous is the impurity (dopant) generally used. In the most common process, a carrier gas is bubbled through phosphorus oxychloride ($POCl_3$), mixed with a small amount of oxygen, and passed down a heated furnace tube in which the wafers are stacked. This grows an oxide layer on the surface of the wafers containing phosphorous (phosphorus glass). At 800°C to 900°C the phosphorous diffuses from the oxide into the silicon. After 20 min, the phosphorous impurities override the boron impurities in the surface region of the wafer and generate a thin, heavily doped n-type region. Sometimes additional boron-implantation is carried out to generate a back-surface-field (BSF) by "emitter-diffusion" (gas phase diffusion at 900°C),

- **Removal of the phosphorus oxide** ("phosphorous glass") via wet-chemical process at 20°C to 60°C, if required, etches away the junctions at the side and back of the cell.

- **Attachment of electrical contacts:** In standard technology the process used is known as "vacuum evaporation". The metal to be deposited, mostly Al, is heated in a vacuum until evaporation. The metal vapor will then condense in the cooler solar cells, creating the contact. For the back contact the metal is normally deposited over the entire back surface. A shadow mask is used for the front contacts to create the typical grid pattern on the top surface of solar cells ("screen printing process"). The material used is mainly Ti/Pd/Ag. After that the contacts are sintered at 400°C in an IR-oven,

- **An optical anti-reflective** layer is created traditionally by sputtering TiO_2 or Ta_2O_5 in vacuum and sintering it at 400°C, nowadays hydrogenated silicon nidride SiNx:H is state-of-the-art, thus lifting cell efficiencies by 1.5%.

- **Testing:** visual control of the cells – checking for homogeneity of the anti-reflective layer. Mechanical test – durability of contact bonding. Opto-electrical test - recording of the current-voltage characteristics at an irradiance equivalent to a spectrum of AM 1.5, classification of the cells into groups of equivalent short-circuit current and form-factor.

Electrical energy consumption for these processes in the first model case: 216 kWh per m^2 of cell area, fuel consumption is 116 kWh/m^2 and non-energetic fuel consumption is 12 kWh/m^2. For the second case model: electrical energy consumption is 102 kWh/m^2, fuel consumption 39 kWh/m^2, and non-energetic fuel consumption remains constant at 12 kWh/m^2. A more detailed derivation of this calculation can be found in the annex.

While many processes, such as gas phase diffusion, show a high base load, actual use of production capacity determines the energy consumption in processing. Processing energy consumption can be reduced to 30% of the original case just by changing from a one shift to a four shift operation mode (Hagedorn 1989). Additional energy consumption by climatization, ventilation, lightning etc. is in the same vicinity as the energy needed for processing.

6.4.7 Multi-Crystalline Wafers to Multi-crystalline Solar Cells

The production of multi-crystalline solar cells is equivalent to the production of single-crystalline solar cells, but requires an extra step: "Passivation" by hydrogen of the grain boundaries of the multi-crystalline material.

- **Cleaning** by etching the oxide in KOH,

- **Doping** (diffusion of emitter): traditionally via a screen printing process with a phosphorous emulsion, then diffusion in an IR-oven at 900°C for maximal 30 minutes (resulting depth of diffusion: 0.3–0.5 µm) – nowadays gas diffusion is often preferred,

- **Removing of phosphorous glass** (wet-chemical),

- **Passivation of grain boundaries:** Processing of the wafers in a hydrogen plasma at 300°C for 30 minutes,

- **Application of contacts:** screen printing (backside Ag/Al paste, front side Ag paste) sintering in an IR-oven (at maximal 700°C for 20 minutes),

- **Generation of an anti-reflective layer** by a screen printing process with a TiO_x-emulsion, sintered at room temperature (20°C),

- **Testing:** Visual cell test: checking homogeneity of the anti-reflective coating on the cell, mechanical tests: contact bonding, opto-electrical cell test: recording of the current-voltage characteristics at an irradiance equivalent to a spectrum of AM 1.5, classification of the cells into groups of equivalent short-circuit current and form-factor.

Energy consumption of these processes is not very different from the one of single-crystalline solar cells: electrical energy consumption for the first mode case amounts 213 kWh per m^2 of cell area, fuel consumption is 107 kWh per m^2, and non-energetic fuel consumption is 8 kWh/m^2.

For the second mode case electrical energy consumption is 89 kWh/m^2 and fuel consumption is 34 kWh/m^2. Non-energetic fuel consumption is 10 kWh/m^2 (Hagedorn 1989). Using wafers of a thickness of only 0.2 mm and the transition to a four shift operation of the factory would reduce primary energy consumption and carbon dioxide emissions by half.

6.4.8 Production of Amorphous Silicon Solar Cells

These cells are made from thin films of amorphous silicon incorporated with a small percentage of hydrogen. This greatly reduces the electrical resistance of the material and allows it to be doped n-type or p-type. The design of the cells optimizes the collection of current by having a very thin n-layer and players with an intrinsic (i) layer thick enough to absorb almost all of the incident light, to give a p-i-n structure. The electric field created by the p-n junction extends throughout the i-layer and greatly aids the collection of the photogenerated charge carriers. Unfortunately the electrical properties of the i-layer are degraded by the absorption of light and the performance of the cells decreases with exposure to sunlight. The effect can be reduced, but not eliminated, by careful control of the deposition conditions, and reduced further by using multiple junctions, each with a thinner i-layer.

The best efficiency so far recorded for an amorphous silicon cell is 13.6% on a triple junction stack.

6.4.9 Production of Solar Cells Made of other Semiconductors

Gallium arsenide (GaAs)
GaAs has a well-developed technology which hinges on the commercial interest for this material for the production of blue light-emitting diodes and semi-conductor lasers. Because of its almost ideal band-gap and its

corresponding spectral efficiency, the most efficient solar cells have been based on GaAs. The limited resources of gallium (production only about 30 tons/a) ensures that GaAs will always be an expensive solar cell material. For concentrator systems, the amount of material required for given power output can be reduced. The direct band gap of GaAs also means that irradiance is absorbed very quickly after entering, therefore only layers of a few microns are required. The big disadvantage is the toxic nature of arsenic and thus environmental consequences of deploying large GaAs solar energy systems would have to be carefully examined.

Table 6.7. Accumulated energy consumption for the production of silicon wafers

Type/ Process	MG-Si in kWh/kg	EG-Si in kWh/kg	Cell-Si in kWh/kg	Sawing in kWh/kg	Yield in m²/kg	References
Single-crystalline Silicon (sc-Si)	n. m.	416	150	5	0.38	Aulich et al. 1986
	n. m.	139 (purest Si)	150	5	0.38	Aulich et al. 1986
	20	200			0.856	Palz et al. 1991
	20	100				Hagedorn 1989a
	14.6	114.3 (-29.8)				Hagedorn 1989
Multi-crystalline Silicon (mc-Si)		416 (139)	25	5	0.38	Aulich et al. 1986 (crucible casting)
		416 (139)	35	<5	> 0.4	Aulich et al. 1986 (EFG)
		416 (139)	7	<5	> 0.4	Aulich et al. 1986 (S-Web)
	20	100				Hagedorn 1989a
	14.6	114.3 (-29.8)				Hagedorn 1989
	20	200		20	0.856	Palz et al. 1991

Cases   I: case one: 2 MW_p/a, single shift operation of production plant
          II: case two: 25 MW_p/a, four shift operation of production plant

Copper Sulfide – cadmium Sulfide Cells (Cu_2S – CdS)

CdS solar cells have a history of development dating back to 1954. Since then there have been several attempts to produce a commercial solar cell based on this material. The striking feature of these cells is the ease by which they can be fabricated. Because fine-grained multi-crystalline CdS is also adequate as a substrate material, numerous options for preparing such substrates are available. Vacuum evaporation and spraying method seem the most promising. The major disadvantages of these cells are their low efficiency and their lack of the inherent stability possessed by silicon solar cells. With low efficiencies, the costs of other parts in the system become more important because the area required for a given output increases. Balance-of-system costs, such as those due to site preparation, support structure, and wiring can dominate PV system costs to such an extent that even if the cells were free it would still be cheaper to use higher-efficiency cells of higher cost. As a rule of thumb, 10% module efficiency is the lowest that can probably be tolerated for cost-effective large-scale generation of PV power (Green 1982).

Distribution of Cell Production by Technology

(Reference: P.D. Maycock, PV market update, 2003)

- Multi-crystalline Silicon: 58%
- Single-crystalline Silicon: 32%
- Thin film technology: 7%
- Others: 3%

6.5 Production of PV Modules

Six to twelve solar cells are connected in series to give a so called "string". The cell connecting tabs consist traditionally of flat silver wires. Contacting is then carried out by point or infrared welding. At present, cooper flat-wire tabs (thickness 100–200 µm) are used, coated by an alloy consisting of 60% Sn, 38% Pb, and 2% Ag. Those tabs are sold by flame, IR, hot air, Laser or eddy current. To avoid the hazardous lead, increasingly a coating alloy of 96.5% and 3.5% Ag is used, but fusion temperature of that is 40 K higher. Three to twelve such strings are the base for a solar module (or PV module), which protects the PV generator against weather and the environment. For that the cell matrix is laminated together with a front sheet of glass, transparent plastic and a backside glass or foil. The most common plastic used for this purpose is a copolymer: Ethylene-vinyl-acetate (EVA). The backside foil consists often of a Tedlar®–Polyester–Tedlar® or Tedlar®–Aluminum–Tedlar® laminate.

A frame made of aluminum, stainless steel or plastic and a terminal box complete the module. The aluminum frame increases the energy requirement for the module production by 215 kWh per m^2 of module area (see e.g. Aulich et al. 1986). For this purpose (and to reduce costs) frameless PV modules are getting more common. Mounting and fixing to the support structure is then usually done by clamping bolts, sometimes a glue fixture is used. Energy requirement and CO_2 emissions are reduced significantly by the use of frameless modules.

Table 6.8. Gross Energy Requirements (GER) of module materials (Alsema 1998)

Material	Density in kg/m³	GER in MJ/kg	Certainty of data	Regional validity	References
Float glass	2,500	15	very good	EU	Gelen 1994 Hantsche 1991
Al primary	2,700	190	very good	EU, US	Worrell 1994
Al secondary	2,700	18	very good	EU, US	Gelen 1994
Al profile, anodized (from primary Al)	2,700	220	very good	EU, US	Keoleian and Lexis 1997 Worrell 1994 Gelen 1994
EVA	900	75	fair	EU	Gelen 1994
PVF (Tedlar)	1,800	115	fair	EU	Gelen 1994

6.5.1 Lamination Process

The compound of the module (glass – EVA – solar cell matrix – EVA – glass, respectively of glass – EVA – solar cell matrix – EVA – backside laminate) is created using pressure and heat. Usually vacuum-laminators are used during the full lamination process (heating, curing of the co-polymer EVA, and cooling). Other methods, such as using a separate oven for curing under normal pressure are in discussion. "Curing" at normal pressure reduces the cycle time in mass production and the laminator can be reloaded quickly. Energy loss is reduced because the laminator does not need to be cooled down. The use of modern "fast cure" EVA as an encapsulate allows for the reduction of the curing phase from 22 minutes to 4 minutes; thus reducing the advantage of separate or "equalized" processing. Discussed later in the text, a vacuum laminator with an integrated heating and cooling system is used. Because of its significance for energy efficiency and possible reduction of CO_2 emissions, the new separate processing is also disucssed within a suggestion of the Austrian company Isovolta AG for a "passing through" laminator.

Presently, the "constant temperature" method at 150°C is increasingly used, controlled laying down and lifting up the modules from the heating plates in the vacuum oven. This method increases manufacturing throughput considerably. In Japan two-stage processing is the method-of-choice. Pre-lamination in a vacuum laminator, then curing in a circulating air oven, avoids "fast-curing" EVA for the cost of additional handling.

In the future, TPU (thermoplastic polyurethane) may play a major role, because it does not require any curing-time, a suitable roller-laminator is under development by the company Bayer. However, the costs for TPU are still double than for standard EVA.

The equipment for the lamination process undergoes fast development – the brand names and values given in the tables below therefore just represent typical examples for the calculation of the specific energy requirements for PV modules.

6.5.1.1 Integrated Laminator

A flexible membrane in the cover plate of the vacuum laminator (which creates a separate chamber), builds up the necessary pressure on the laminate (heated by a heating plate from below) by being applied to ambient pressure on its upper side (release of vacuum). Procedure of the lamination process as suggested by Isovolta for conventional EVA:

1. Load of laminate (90°C), create a vacuum of ca. 10 mbar,
2. Heat up from 90°C to 155°C within 10 minutes ($\triangle$ 6,5 K/min),

3. At 120°C membrane pressure is created by applying ambient pressure to the top plate.
4. After reaching 155°C, maintain temperature for 15 minutes,
5. Cool-down from 155°C to 90°C within 10 min (≙ -6.5 K/min),
6. After reaching 100°C open laminator.

The procedure of the lamination process according to the "Springborn" method (see Photocap 1996) with conventional and "fast cure" EVA:

1. Pre-heat the heating plates of the laminator to 75°C,
2. Attach two silicon or a Teflon® separator foil, load the laminator,
3. Create a vacuum in the bottom of the laminator (< 1.3 mbar, reached after 3 minutes),
4. Build-up pressure (1 bar) in the cover plate of the laminator, as soon as the laminate reaches the temperature of 60°C (also after 3 minutes), heat up further,
5. Curing (from 8 minutes after loading):
   For normal EVA at 155°C: 22 minutes,
   For "fast cure"-EVA at 145°C: 4 minutes
6. Open, cool and carefully unload (EVA is still soft),
7. Set temperature to 75°C for the next cycle.

Major differences to the Isovolta processing is a steeper time-temperature ramp by 15 K/min (instead of 6.5 K/min) and pressurization at 60°C (instead of 120°C), before the melting point of EVA at 80°C.

To compare primary energy consumption of heating, electrical energy generation for the heating plates is considered to be carried out at an efficiency of 35%. Losses via heat conduction to the containment and convection of the evacuated laminated are neglected.

Due to the good thermal insulation of the heating-mattress towards the ambient, power consumption consists mainly of the heat-flow from the heater into the laminate and the electrical heating-mattress itself. A big (3 m^2), thin (5 mm) laminate of glass, EVA and solar cells needs about 1 kWh to be heated from 25°C to 150°C:

$$Q_{Laminate} = c_p \cdot \rho \cdot V (155°C - 20°C) = 1.164 \text{ kWh}$$

using c_p = 0.23 Wh/(kg K) and ρ = 2,500 kg/m^3 (for glass).

A heating power of 8.73 kW is necessary, if the final temperature should be reached within 8 minutes. For the heating mattress, which weights 37.5 kg with an $c_p = 0.108$ Wh/(kg K) for copper, another 0.547 kWh have to be added, so a total power of 12.83 kW is required. Cooling power is equivalent or lower, while the laminate could be removed at 90°C. If a cooling system based on tubes is used (instead of air ventilation), its heat capacity has to be considered also. It is taken for granted that 10% of total power consumption is required to keep the temperature to 155°C.

6.5.1.2 "Passing-Through" Laminator

Temperature cycles for the "passing-through" process are the same as for the process using the integrated laminator. The laminate is transported by a conveyor belt through the vacuum laminator, the curing oven and the cooling zone. Cycle time is ten minutes.

Isovolta gives a power consumption of 30 kW for the whole system, while 9 kW are required for the vacuum laminator (up to 1.2 m² laminates here), 12 kW for the curing oven double the size, and the rest for cooling ventilation, process control and handling. According to Springborn, the laminate requires three minutes of pumping to be without air bubbles and can be pressurized as the temperature reaches 60°C, at the same time the power for heating is increased and kept at 155°C. The data of Table 17 and 18 shows, how simultaneous lamination of several modules reduces energy consumption significantly (comparison of ICOLAM II from Isovolta with SPI 460 from Spire). For all types of laminators the energy consumption per cycle was calculated for an eight minute heating period at maximum power. Power consumption to keep the temperature over 4 resp. 22 min was considered by using 10% of the maximum power value. The cycle time of the laminator for normal EVA can be equalized to "fast cure" EVA, if an additional "curing oven" is used, but adding an additional 10 kW for power requirement. Due to the short cycle time of 4 minutes, the integrated laminator using "fast cure" EVA has the lowest energy consumption. According to the data supplied by the manufacturer and the estimations given above in tables the separation of the lamination, curing and cooling processes at the ICOLAM II do not have a positive effect on the reduction of energy flows [18]. Lamination by conventional laminators cause primary

---

[18] To reduce energy losses by changing the temperature of the heating plate, two manufactures of laminators since 1996 (NPC and S.E. Project: 26 kW for four 83.6 W_p modules) keep the temperature of the hearing plates constant at 155°C at initiate the cooling process by lifting the laminate off the hearing plate. According to the manufacturers energy consumption is only 20% of conventional laminators. More reasonable sees a saving of 30% in accordance to the relation of the heat capacity of the laminate to the heat capacity of the hearing plate.

energy consumption of 3 kWh/m². Using active cooling, 5 kWh/m² are a realistic value.

Table 6.9. Power- and energy requirements of lamination with square 10 cm cells

Type of laminator	Length x width in cm	Power in kW	Water in l/cycle	Cycle in min	No. of mod. per cycle	Output mod./a	Output MW_p/a	El. energy cons. in kWh/kW_p
SPI 460	200 x 137	165	30	33	3	10,800	0.78	12.98
SPI 460, fast cure	200 x 137	165	30	15	3	23,760	1.71	11.71
ICOLAM II	80 x 150	30	40	10	1	12,000	0.86	55.56
Solarwerk	230 x 150	20	0	33	4	17,600	1.15	11.81
Solarwerk fast cure	230 x 150	20	0	15	4	38,720	4.32	9.73

Type of module: PL-800, module size: 56 cm x 122 cm, power: 72 W_p; 3 min for load and unload at an integrated laminator; 2,000 h/a

Table 6.10. Power- and energy requirements of lamination with square 15 cm cells

Type of laminator	Length x width in cm	Power in kW	Cycle in min	Modules per cycle	Output: modules/a	Output: MW_p/a	El. energy consum. in kWh/kW_p
SPI 460	200 x 137	16.5	33	2	7,200	0.622	16.23
SPI 460, fast cure	200 x137	16.5	15	2	15,840	13.68	13.37
ICOLAM II	80 x150	30.0	10	1	12,000	1.037	46.30
Solarwerk	230 x150	20.0	33	3	10,800	0.933	13.12
Solarwerk fast cure	230 x150	20.0	15	3	23,760	2.053	10.80

Module: size: 67.2 cm ·146.4 cm; power: 86.4 W_p; 3 min for load/unload at the integrated laminator; 2,000 h/a, water use equiv. to Table 6.9.

6.5.2 Production of "Encapsulated" PV Modules

An "encapsulated" PV module consists of the following compound:

glass sheet – plastic – solar cell matrix – plastic – glass sheet

Most often, EVA (ethylene vinyl acetate) is used as the plastic layer which a allows for a tight-fit of the glass sheets and cell matrix. To avoid air bubbles in the laminate, a vacuum laminator is used to pressurize the compound during the lamination process at 145°C to 200°C. The copolymer EVA, supplied in foils at a thickness of 0.5–0.7 mm, "cures" during the lamination, making its melting process irreversible. To prevent the plastic from UV degradation ("yellowing" or "browning") UV absorbers are added to the EVA. The glass sheets consist of thermally tempered, iron-free ("white") glass of a thickness of 2–3 mm.

This kind of module production has, for the first model case: a need for 81 kWh per m^2 of cell area of electrical energy consumption, 80 kWh per m^2 of combustible fuel consumption and 20 kWh per m^2 of non-energetic consumption. The second model case: electrical energy consumption of 36 kWh per m^2, combustible fuel consumption of 64 kWh and 13 kWh non-energetic fuel consumption. Exact values for the different components are given in Table A5.

6.5.3 Production of "Laminated" PV Modules

A "laminated" PV module consists of the following compound:

glass sheet – plastic – solar cell matrix – plastic – plastic foil compound

Most often EVA (ethylene vinyl acetate) is used as plastic. Opposite to the encapsulated module, the backside is not consisting of a glass sheet, but of a Tedlar®–Polyester–Tedlar® or a Tedlar®–Aluminum–Tedlar® foil compound with a thickness of 0.5 mm. The equivalent energy requirements are given in Table 6.8. Aulich et al. 1986 is giving 27 kWh/m^2 for a strengthened, iron-free glass (3 to 4 mm of thickness); for the plastic 23 kWh/m^2, and for the lamination process 5 kWh/m^2. An extensive overview for the energy consumption values of the different materials is shown in Table A5 in the Annex.

Energy Consumption for the Set-Up of a PV Power Plant   111

Table 6.11. Cumulated energy consumption for the production of PV modules

Type	Wafer prod. in kWh/m²	Cell prod. in kWh/m²	Laminate prod. in kWh/m²	Module framing in kWh/m²	Heating, light of factory in kWh/m²	Sum of energy in kWh/m²	Reference
Single-cryst.-Si-module	780	26	55	215	-	1076 / 861[1]	Aulich et al. 1986
	470	126 (I) / 110 (II)	57	0	-	653 / 637	Hagedorn 1989a
	-	24	57	0	46	-	Palz et al. 1991
		344 (I) / 153 (II)					Hagedorn 1989
Multi-cryst.-Si-module	200	26	55	215	-	496 / 281[1]	Aulich et al. 1986
	220	24	57	0	46	347	Palz et al. 1991
	220 (I) / 215 (II)	126 (I) / 110 (II)	47	0	incl. / incl.	393 (I) / 372 (II)	Hagedorn 1989a
		328 (I) / 133 (II)					Hagedorn 1989
Encapsulated PV module				181 (I) / 113 (II)			
Laminated PV module			55				

[1] without frame

(I) first case: 2 MW$_p$/a, single shift operation
(II) second case: 25 MW$_p$/a, four shift operation

6.5.4 Electrical Power Conditioning

To adapt current and voltage from the PV generator to the load requirements, different kind of converters may be necessary depending on the type of system. For an autonomous system a charge controller, deep discharge prevention and the electrochemical storage are needed. Also favorable is a Maximum-Power-Point-Tracker (MPPT) for optimal adaptation of the load to the actual current-voltage characteristics of the PV generator and a DC-AC converter (inverter), which allows the use of AC loads.

PV systems for electrical grid injection require an inverter capable to synchronize to the grid, a Maximum-Power-Point-Tracker is also favorable. Very often these two devices are combined. The energetic requirements for the units, including wiring, for different scales of power plants are given in Table 6.12. A detailed listing of all materials involved in power conditioning components are given in Table A16 in the Annex. The power generation composition and specific emissions of the producer countries of some raw materials are given in Table A18.

Table 6.12. Cumulated primary energy consumption for components for grid injection

PV system type	Cell type	Connection, wiring etc. in kWh/kW$_p$	Inverter in kWh/kW$_p$	References
Large scale power plants (1500 kW$_p$)	sc-Si	110	176	Hagedorn 1989
	a-Si	146	176	Hagedorn 1989
Small scale power plants (300 kW$_p$)	sc-Si	525	352	Hagedorn 1989
	a-Si	755	352	Hagedorn 1989
Residential roof top systems (2.5 kW$_p$)	-	-	222–533	Johnson et al. 1997

sc: single-crystalline, a: amorphous, Si: Silicon
Minimal value given at Johnson et al. 1997 is for a recycling quote of 50%

6.5.5 Support Structure

In order to mount photovoltaic modules securely at the installation site, foundations and support structure have to be able to withstand the maximum expected wind speed during lifetime. Table 6.13 shows different types of support structures and the according energy requirements.

Table 6.13. Cummulated energy consumption for PV support structures

Type of support structure	Steel in kg/m^2	Concrete in kg/m^2	Energy requirements in kWh(el)/m^2	References
Roof mount	< 10	0	< 75	Bloss et al. 1992 Hagedorn et al. 1992
	< 10	0	< 30	Winter et al. 1986
Small scale (e.g. Pellworm)	13.3	67	100	Hagedorn et al. 1992
			45	Winter et al. 1986
Large scale (e.g. Kobern-Gondorf)	24.6	115.2	170	Hagedorn et al. 1992
			82	Winter et al. 1986
	235 kg/kW$_p$	1,896 kg/kW$_p$		Muller et al. 1997
Two axis tracking	46.3	980	450	Hagedorn et al. 1992
			205	Winter et al. 1986
Two axis tracking (series production USA)	14 – 26	ca. 200	186	Hagedorn et al. 1992
			90	Winter et al. 1986
Comments	\multicolumn{4}{l}{Values for the energy requirements at tracking systems and roof mounting at Hagedorn et al. 1992 have been extrapolated}

6.6 Installation and Taking into Operation

The energy requirement for installation at the installation site (W_{setup}) consists of the transport requirements for all components and tools ($W_{transport}$), the support requirement, inclusive land- and foundation preparations (W_{mount}) – see also Tables 6.13 and 6.14 –, and also losses caused by subnormal operation during checking and adjustment (W_{adjust}). Aside from the PV power plant itself, also the personnel, tools and consumption materials have to be taken into account.

$$W_{setup} = \sum W_{transport} + \sum W_{mount} + \sum W_{adjust} \qquad (12)$$

Table 6.14. Cummulated energy consumption and CO_2 emissions for construction work

Tool	Direct energy consumption in MJ/m^3	Total energy requirements in MJ/m^3	Direct CO_2 emissions in kg/m^3	Total CO_2 emissions in kg/m^3
Caterpillar	5.9	7.81	0.0107	0.5324
Hydraulic excavator	4.9	6.71	0.00938	0.4550

Comments: values related to excavated material (Frischknecht et al. 1996)

6.6.1 Transport

The power plant itself, the personnel and the tools have to be transported to the site of installation. The energy requirements are given in Table 6.3 and in Table A8 in the Annex.

6.6.2 Installation

To compute the primary energy consumption for installation (W_{mount}) the typical need for materials, tools, and energy has to be considered. The according data is given in the Annex in the Tables A4 and A5.

6.6.3 Setting into Operation

During the adjustment-, calibration-, and test phase a reduced energy yield is well possible. Sometimes components have to be exchanged or added. These requirements are considered by W_{adjust}.

6.7 Operation Expenses

For operation (W_{op}) as well as direct expenses such as cleaning of the module surface (W_{clean}) and preservation of the support structure ($W_{maintain}$), unexpected repairs, such as those caused by lightning or earthquakes or vandalism, have also to be considered (W_{repair}). Sometimes, the potential for alternative use of the area has to be compensated for, e.g., as for agriculture (W_{area}). If conventional building structure is substituted by the PV installation, e.g., in case of a solar roof, a negative value has to be used as a cost input.

$$W_{op} = \sum W_{clean} + \sum W_{maintain} + \sum W_{repair} + \sum W_{area} \qquad (13)$$

6.7.1 Cleaning

The front surface of the PV modules will accumulate blowing dirt or excrements from birds which will reduce the yield; therefore, on-site monitoring and cleaning are necessary. This creates supervision costs (W_{test}), personnel, transport ($W_{transport}$), cleaning media ($W_{utilities}$) and, in case of mechanized cleaning processing, as, e.g., vapor spray machinery, the delivery of energy for its operation ($W_{cl.process}$). Usually a cleaning interval of a year is practical.

$$W_{clean} = \sum W_{transport} + \sum W_{test} + \sum W_{cl.process} + \sum W_{utilities} \qquad (14)$$

6.7.2 Maintenance

The support structure has to be protected from corrosion. If non-corrosive materials are not used, a durable surface treatment (e.g., varnish) has to be applied and renewed on-site. This leads to costs for personnel, transport (approach and departure) ($W_{transport}$) and tools (W_{tools}). If a battery storage is used, the batteries have to be maintained or even exchanged ($W_{materials}$).

$$W_{maintain} = \sum W_{transport} + \sum W_{tools} + \sum W_{process} + \sum W_{materials} \qquad (15)$$

If components fail, they have to be exchanged on-site. Hereby transport of personnel and material ($W_{transport}$), the determination of damage (W_{test}), the tools (W_{tools}), the materials themselves ($W_{materials}$) and the energy consumption ($W_{rep.process}$) on-site are to be considered.

$$W_{repair} = \sum W_{transport} + \sum W_{test} + \sum W_{rep.process} + \sum W_{tools} + \sum W_{materials} \quad (16)$$

6.7.3 Use of Land

The area used for the operation of the power plant has, under circumstances, to be purchased, rented or leased. By further use as agricultural area the effects caused by the shadowing have to be considered in an extra energy- and CO_2- balance (W_{area}). At most of the cases the area covered by PV could be fully used (roof mounting), or even some building components can be substituted (facade- and roof integration). To avoid yield losses by shadowing in dense built-up area, additional land in front of the PV array has to be reserved (with no obstacles on it).

6.8 Dismantling

At the end of its useable lifetime the installation has to be dismantled ($W_{deconst}$), transported ($W_{transport}$), dumped (W_{waste}) or recycled ($W_{recycle}$). If the site will not be used for a new installation, the place of installation has to be restored ($W_{reconstsite}$).

$$W_{de} = \sum W_{deconst} + \sum W_{transport} + \sum W_{reconst\,site} \pm \sum W_{recycle} \mp \sum W_{waste} \quad (17)$$

6.8.1 Dismantling

Dismantling includes transport of personnel and tools to the site and direct energy necessities. Levin 1993 uses the same values as for installation. For total investment costs of 7,500 US/W_p installation contributes 4.6%. For primary energy use, Levin 1993 calculates 43 kWh/kW$_p$ (the same as for installation minus the materials used). The data for different constructions and services are available in the Tables 6.12 to 6.14 and A4 to A5, but the values are depending on the individual construction of each PV power plant.

6.8.2 Transport

This includes transport from the installation site to the location for recycling. Here the values of Table 6.3 and A8 are helpful. Levin 1993 calculates 107 kWh/kW$_p$ for transport from the installation site. The energy gain by recycling will be considered in the following chapter.

7
Energy Yield

Within a Ph.D. thesis (Krauter 1993c) an extensive operational model was developed in order to quantify losses caused by poor optical and thermal adaption of solar-electrical converters to real operating conditions. That model, together with well-known electrical performance modeling, was used as a platform and tool to determine, as accurately as possible, the yield of PV systems.

The modeling of the PV system was carried out in such a way that all parameters that had an effect on the electrical yield (and therefore on the specific CO_2 production) of more than 1% are taken into consideration.

7.1 Model to Determine the Cell Reaching Irradiance

The path of irradiance from the Sun, through the Earth's atmosphere and through the solar panel mechanism of photovoltaic power conversion, the spectrum to actually reach the cell is extracted.

7.1.1 Sun's Position Relative to Earth's Surface

Previous versions of the model (Krauter 1993c, Strauß 1994) used the equations given by 1978, considering follow-up reports and improvements by Archer 1980, Wilkinson 1983, and Kambezidis 1990. These models don't work in situations when the Sun's elevation is greater than 90° (occurring for locations between the Tropic of Capricorn and the Tropic of Cancer – between 23.5° South and 23.5° North of the Equator). These parts of the equations have been substituted by the equivalent formulas given by in DIN 5034 part 2. According to DIN 5034 part 2, the actual position of the Sun, determined by the elevation angle of the Sun γ_s and the azimuth angle of the Sun α_s (see Fig. 7.1), can be calculated as:

$$\alpha_s = 180° - \arccos\left(\sin(\gamma_s)\cdot\sin(\phi) - \frac{\sin(\delta)}{\cos(\gamma_s)\cdot\cos(\phi)}\right) \qquad (18)$$

$$\gamma_s = \arcsin(\cos((12-t)\cdot 15°)\cdot \cos(\phi)\cdot \cos(\delta) + \sin((12-t)\cdot 15°)\cdot \sin(\delta)) \quad (19)$$

$$\delta = 0.3948 - 23.2559\cos\left(\frac{d\cdot 360°}{365} + 9.1°\right)$$
$$- 0.3915\cos\left(\frac{2d\cdot 360°}{365} + 5.4°\right) - 0.1764\cos\left(\frac{3d\cdot 360°}{365} + 26°\right) \quad (20)$$

- d number of days during a year, starting with $d=1$ for January 1st (no leap year)
- t solar local time (highest position of sun at noon, 24 hour system)
- ϕ Latitude (positive value for locations North of the equator, negative value for locations south of the Equator)
- δ Declination, position of the Sun at solar noon (highest elevation during a day) relative to the Equator

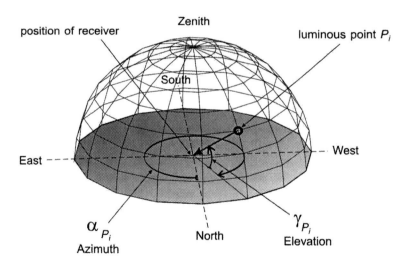

Fig. 7.1. Definition of the elevation angle γ_s and the azimuth angle α_s of a luminous point P_i (e.g,. the Sun) according to DIN 5034 part 2.

The formulas have been modified to take into account the refractive properties of the atmosphere, depending on air pressure and temperature, as well as the bow of the Earth's surface. The Astronomical Almanac 1996 is suggesting the following corrections:

for elevation angles of the Sun $\gamma_S \geq 15°$:

$$\Delta\gamma_S = \frac{0.00452° \cdot p}{T_A \tan\gamma_S} \tag{21}$$

for elevation angles of the Sun $\gamma_S < 15°$:

$$\Delta\gamma_S = \frac{(0.1594 + 0.0196\,\gamma_S + 0.00002\,\gamma_S^2) \cdot p}{(1 + 0.505\,\gamma_S + 0.0845\,\gamma_S^2) \cdot T_A} \tag{22}$$

using:

$\Delta\gamma_S$ Deviation of the Sun's elevation angle due to atmospheric refraction in angular degrees (additional to the real elevation angle of the Sun γ_S)

γ_S Real elevation angle of the Sun in angular degrees

p Atmospheric air pressure in mbar

T_A Ambient temperature in K

At the transition point, for an elevation angle of 15°, only small differences of the two functions and the derivations occur, which results in a smooth and steady course of the total function which considers atmospheric refraction. The accuracy is about 0.0017° for $\gamma_S \geq 15°$. The incidence angle θ_{in} onto the PV module surface is determined by the position of the Sun (γ_S, α_S), by the module elevation γ_M, and by the module azimuth α_M as follows:

$$\theta_{in} = \arccos\left(\sin\gamma_S \cdot \cos\gamma_M - \cos\gamma_S \cdot \sin\gamma_M \cdot (\alpha_S - \alpha_M)\right) \tag{23}$$

Figure 7.2 shows the results of calculations of Sun's elevation, Sun's azimuth, and the incidence angle onto a solar module (module elevation angle 36°, directed southward) during the course of a summer day (6/21 at 36°N).

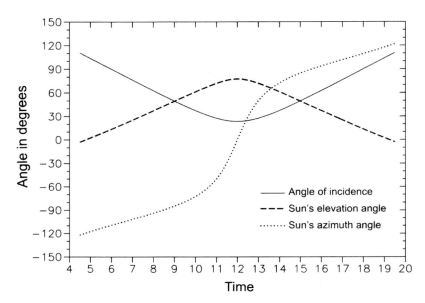

Fig. 7.2. Sun's elevation γ_S, Sun's azimuth α_S, and incidence angle θ_{in} onto module surface during a day (6/21, lat=ϕ=36°N, γ_M=36° directed towards South).

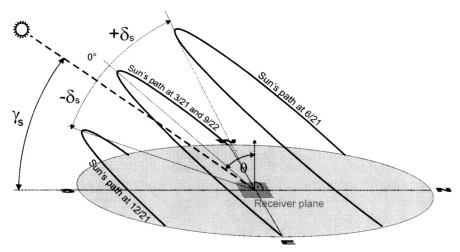

Fig. 7.3. Path of the sun on the northern hemisphere at Equinox (3/21 and 9/22), summer (6/21) and winter (21/12). Shown are also maximal deviations δ_S of the midday points and the actual sun's elevation angle γ_S, with the corresponding incidence angle θ on a receiver (after Ertürk 1997).

7.1.2 Way of Sun's Irradiance Through the Earth's Atmosphere

7.1.2.1 Solar Constant

Directly above the Earth's atmosphere the Sun's irradiance amounts to 1,353 ± 7 W/m². This value is called "Solar Constant" E_0. The distance from the Earth to the Sun does not remain absolutely constant, due the slightly elliptic course, thus the solar constant is varying a little bit (< 1%).

7.1.2.2 Global Irradiance

Earth's atmosphere absorbs and scatters a part of the irradiance incindencing. Absorption and the probability of scattering depend on the angle of incidence and on the wavelength of solar irradiance. The scattered component is not entirely lost, a part of it will reach the Earth's surface as so called diffuse irradiance. The sum of direct and diffuse irradiance is called global irradiance. Table 7.1 shows reference values for global irradiance at clear sky conditions as a function of sun elevation angle γ_S.

Table 7.1. Global irradiance as function of Sun's elevation angle γ_S (Holman 1990)

Elevation angle of the Sun γ_S	Irradiance E_{glob} in W/m²
5°	41.9
10°	112.8
20°	290.7
30°	472.1
40°	636.0
50°	781.4
60°	901.1
70°	991.8
80°	1,043
90°	1,063

Daily sums of global irradiance over a year for different orientations of the receiver, respectively the PV module, are shown in Figure 50 for a central European location (Krefeld, Germany, 51° N, 11.5° E).

122   Solar Electric Power Generation

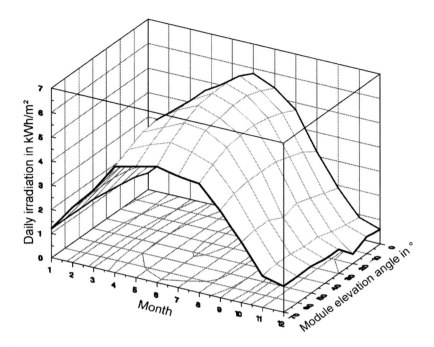

Fig. 7.4. Daily global irradiation over a year as a function of elevation angle of the receiver plane for a location in Central Europe (51° N, 11.5° E), data Schoedel 1993.

The direct and diffuse components of global irradiance during a clear day can be found in Fig. 7.4.

The share of diffuse in relation to direct irradiance as a function of the relation of global irradiance to solar constant (in other words: the "clarity" of the atmosphere) is presented in Fig. 7.5.

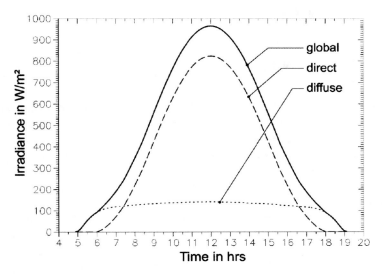

Fig. 7.5. Direct and diffuse components of global irradiance for clear sky conditions at 3/21 (resp. 9/22) for a receiver plane with the same elevation angle as the latitude angle (e.g., 36°), so sun incidents perpendicularly at noon.

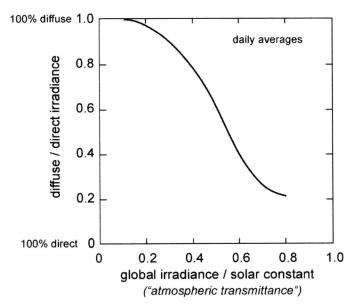

Fig. 7.6. Relative diffuse irradiance component as a function of the relative global irradiance.

7.1.2.3 Direct Irradiance

Direct irradiance is determined by the position of the sun, the air mass, the condition of the atmosphere (temperature, humidity, clouds, aerosols) and the angle of incidence of the sun's radiation towards the receiver plane (e.g., a solar module).

Apart from variation of irradiance changes in spectral composition also occur, which cause changes in photovoltaic conversion efficiency of the solar generator due to its spectral selective sensivity. Considerations for these changes during days and seasons have been carried out just recently (Krauter 1993c etc.).

The CIE spectrum publication No. 85 (1990) was used as a reference for modeling, as it is the state-of-the-art approximation to real conditions and provides data for different air masses (see Fig. 7.7) and turbidity factors. Also, spectra for different H_2O-, CO_2- and O_3-contents were included.

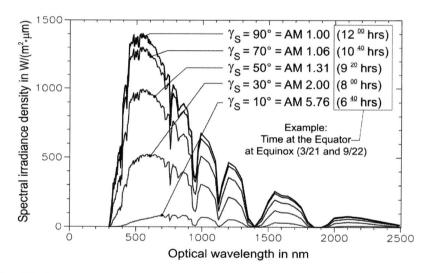

Fig. 7.7. Direct terrestrial irradiance spectra of the sun (AM 1-5.6) according to tables of CIE publication No. 85 (1990) for AM 1, AM 2 and 5.6, other values by interpolation.

7.1.2.4 Diffuse Irradiance

For terrestrial application there is always a considerable diffuse component of the solar irradiance even for clear sky conditions. Over a year the diffuse component amounts to 30% to 60% of the total irradiance (global irradiance). The spatial distribution within the sky's hemisphere of the diffuse or scattered radiation is not homogenous. For an unscattered sun ray, the probability to hit an air molecule or aerosol, and to be reflected, is a function of the air density and thickness of the atmosphere passed through. The reflected component itself is a function of the angle of incidence and the optical refractive indices of the media involved (Fresnel's Law, see chapter 5.1.3.2). Additionally, a multiple scattering has to be considered. A model used for the angular sky irradiance as function of the position of the sun is DIN 5034 part 2 and is given below. An example of the distribution of the sky illuminosity at an elevation angle of the sun of $\gamma_S = 30°$ is given in Figure 54 as a contour plot (projection of the sky's hemisphere to the plane).

$$\frac{L_{eP}}{L_{eZ}} = \frac{\left(1-\exp\left(-\frac{0.32}{\cos\varepsilon}\right)\right)\cdot\left(0.856+16\exp\left(-\frac{3\eta}{rad}\right)+0.3\cos^2\eta\right)}{0.27385\cdot\left(0.856+16\exp\left(-3\left(\frac{\pi}{2}-\frac{\gamma_S}{rad}\right)\right)+0.3\cos^2\left(\frac{\pi}{2}-\frac{\gamma_S}{rad}\right)\right)} \quad (24)$$

$$\frac{E_{eH}}{L_{eZ}} = 7.6752 + 6.1096\cdot 10^{-2}\cdot\gamma_S - 5.9344\cdot 10^{-4}\cdot\gamma_S^2 - 1.6018\cdot 10^{-4}\cdot\gamma_S^3$$
$$+ 3.8082\cdot 10^{-6}\cdot\gamma_S^4 - 3.3126\cdot 10^{-8}\cdot\gamma_S^5 + 1.0343\cdot 10^{-10}\cdot\gamma_S^6 \quad (25)$$

$$E_{eH} = 0.5\cdot E_{e0}\cdot\sin\gamma_S(\tau_a^m - \exp(-T_L\cdot\bar\delta_R\cdot m\cdot\frac{p}{p_0})) \quad (26)$$

$$\tau_a^m = (0.506 - 1.0788\cdot 10^{-2}\cdot T_L)\cdot(1.294 + 2.4417\cdot 10^{-2}\cdot\gamma_S - 3.973\cdot 10^{-4}\cdot\gamma_S^2$$
$$+ 3.8034\cdot 10^{-6}\cdot\gamma_S^3 - 2.2145\cdot 10^{-8}\cdot\gamma_S^4 + 5.8332\cdot 10^{-11}\cdot\gamma_S^5) \quad (27)$$

$$\bar\delta_R\cdot m = \frac{1}{0.9 + 9.4\cdot\sin\gamma_S} \quad (28)$$

L_{eP} Luminosity of a point P at the sky hemisphere
L_{eZ} Luminosity of the zenith
T_L Turbidity factor (function of the state of the atmosphere)
ε angle between Zenith and point P (in °)
η angle between Sun and point P (in °):
   $\eta = \arccos(\sin\gamma_S\cdot\cos\varepsilon + \cos\gamma_S\cdot\sin\varepsilon\cdot\cos(\alpha_S-\alpha_P))$
γ_S Elevation angle of the Sun (in °)
α_S Azimuth angle of the Sun (in °)

α_P Azimuth angle of point P (in °)

$\delta_R \cdot m$ Product of the average optical density (of a pure, dry Rayleigh-atmosphere) and the relative optical air mass m.

Examples for average monthly turbidity factors T_L in Germany are given in Table 7.2.

Table 7.2. Average monthly turbidity factors T_L in Germany (DIN 5034 part 2)

Month	Monthly average of T_L		
	highest	average	lowest
January	4.80	3.8 ± 1.0	3.20
February	4.60	4.2 ± 1.1	3.60
March	5.40	4.8 ± 1.5	4.30
April	5.70	5.2 ± 1.8	4.80
May	5.80	5.4 ± 1.7	4.90
June	7.40	6.4 ± 1.9	5.60
July	6.90	6.3 ± 2.0	5.70
August	6.90	6.1 ± 1.9	5.70
September	6.00	5.5 ± 1.6	5.20
October	4.90	4.3 ± 1.3	4.00
November	4.20	3.7 ± 0.8	3.30
December	4.10	3.6 ± 0.9	3.30
Yearly average	5.40	4.9 ± 1.5	4.70

Spectral modeling of diffuse irradiance is based on the CIE publication No. 85 (1990), analogous to the modeling of the direct component. Figure 54 shows the spectra of the diffuse irradiance for AM 1, AM 1.5, AM 2 and for AM 5.6 under clear sky conditions.

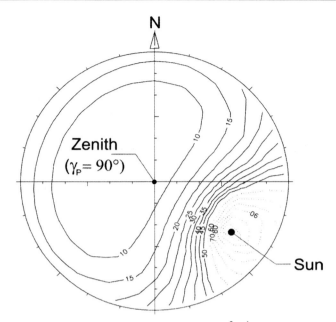

Fig. 7.8. Distribution of illuminousity $L_p(\gamma_P, \alpha_P)$ in W m^{-2}sr^{-1} at the sky hemisphere, calculated according to DIN 5034 part 2 ($\alpha_S=120°$; $\gamma_S=30°$; Zenith in center; North upwards).

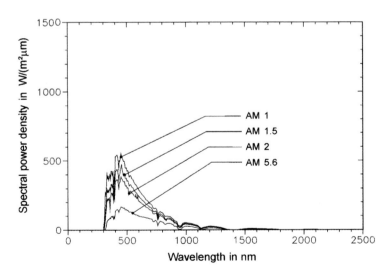

Fig. 7.9. Spectra of diffuse irradiance for different air masses (AM), according to CIE Publication No. 85 (1990).

7.1.2.5 Albedo

A part of the irradiance reaching the ground can be reflected toward the receiver. This so called albedo (-radiation) depends very much on the ground surface properties and incidence angle. While it is very difficult to calculate the albedo, it is not considered in the calculations, if not mentioned differently. For further calculations Table 7.3 is giving some data about the relative amount of the albedo.

Table 7.3. Albedo of different ground surfaces (according to Holman 1990)

Ground surface	Albedo
Dark soil (humid)	0.08
Dark soil (dry)	0.14
Grey ground (dry)	0.25–0.30
Grey ground (humid)	0.10–0.12
Yellow sand	0.35
White sand	0.34–0.40
Rock	0.12–0.15
Grass	0.26
Rice field	0.12
Oak forest	0.18
Water	0.03–0.40
Snow	0.40–0.85

7.1.2.6 Angular Distribution of Yearly Irradiance in Central Europe

The observations shown above use a clear sky as a prerequisite. In reality, the sky over many locations (as e.g., central Europe) is often covered by clouds. Calculations of the angular distribution for Freiburg, Germany (48° N) on the basis of a diffuse-direct distribution of the so called "Typical Meteorological Year" (TMY) using an albedo of 0.2 have been carried by Preu et al. 1995. Figure 7.10 shows the global irradiance as a function of the incidence angle for different module orientations (horizontal and tilted south with module elevation angles of 30°, 48° and 90°). For all incidence angles different from zero, the angular range considered for the cumulative irradiance was one degree. For an incidence angle of exactly 0° the value zero was used, because the angular range considered is infinitely small and so the cumulative irradiance.

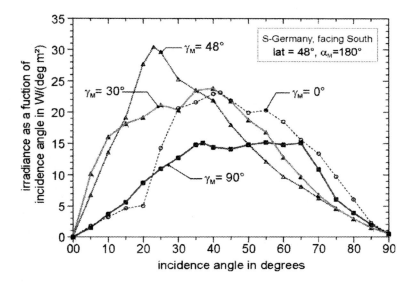

Fig. 7.10. Irradiance as a function of incidence angle for different elevation angles γ_M of the PV module at Freiburg, Germany (48° N), according to Preu et al. 1995.

7.1.3 Optical Model of Module Encapsulation

The optical model described here is mathematically exact for slabs thicker than the wavelength of the irradiance, all possible internal and external reflections are considered. It can be applied for any kind of optical system consisting of a number of different homogeneous plane slabs described by their refractive indexes, absorption coefficients and thicknesses. Optical dispersion and complex refractive indices are included also. Thicknesses could vary from infinite down to the range of the wavelength of the irradiance. In order to calculate solar reflection performance of an optical system during daylight, an irradiance model, considering spatial distribution of direct and diffuse irradiance for clear sky conditions, was implemented.

Reflective losses under realistic conditions
In former contributions estimations of reflective losses at the surface of PV modules have been based on perpendicular incidence and amount to 2-4% of the incoming irradiance. This is correct only for tracking systems without any diffuse parts of insolation (like in space). At non-tracking terrestrial systems direct insolation hits only twice a year really perpendicular onto the surface of the module. At other times the reflected part increases according

to the FRESNEL laws[1] (see below). For terrestrial application there is always a diffuse part of insolation (even at very clear days), amounting during a year to an average of 30% to 60% (depending on the geographic conditions) of the total insolation. An example for the distribution of polarization of diffuse radiation at the sky hemisphere is given in Figure 7.11. This diffuse radiation in anisotropic, the model used later was adopted from DIN 5034, a more advanced model by Perez et al. (1993) will be implemented in the future.

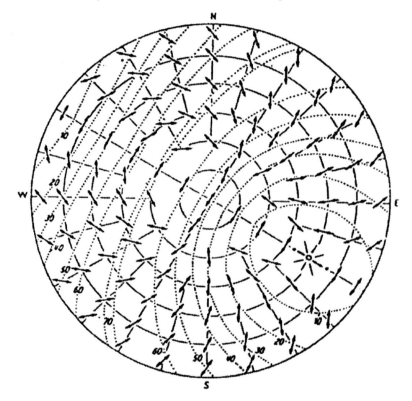

Fig. 7.11. Polarization pattern of the sky hemisphere for a clear day for an elevation angle of the sun of 30° (after von Frisch 1965).

In order to achieve a precise representation of the actual optical conditions in the module, a model for the encapsulation of the cell was developed that extracts the insolation reaching the cell from sun and sky irradiance. This was done by modeling the optical processes (reflection, absorption) happening outside and inside the encapsulation (see also Fig. 7.12). Since diffuse irradiance is scattered, it is also polarized. Therefore reflectance depends on polarization that effect was thought of in former contributions (Krauter et al. 1991, Krauter 1993): Reflection loss increased by 0.5% to 5% over the diffuse part due to polarization. The total amount of reflection losses over a year is in the 20% range (Krauter *et al.* 1994, Krauter and Hanitsch 1996). Reduction of reflective losses to improve transmittance of PV module encapsulation layers provides an important contribution for optimizing PV systems.

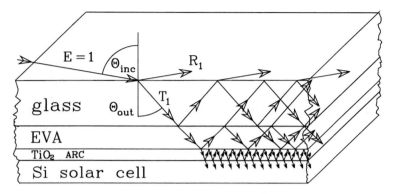

Fig. 7.12. Ray-tracing through a PV module encapsulation.

7.1.3.1 Optical interface at boundary layers

Perpendicular incidences: At the transition when radiation from a material of a given optical density (n_0) enters into another (n_1), the radiation is split up into a reflected component (R) and a transmitted one (T) at the optical interface, see Fig. 7.13 for perpendicular incidence.

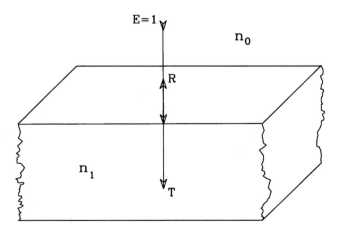

Fig. 7.13. Perpendicular incidence on a plane surface ($\theta_{in} = 0°$).

R is also called "reflectance" and T, "transmittance" (see, e.g., Dietz 1991). Irradiance (E) of the incoming radiation is normalized to $E = 1$.

$$R = \frac{(n_1 - n_0)^2}{(n_0 + n_1)^2} \tag{29}$$

$$T = 1 - R \tag{30}$$

Non-perpendicular incidences: More realistic is the case of non-perpendicular incidences of insolation (see Fig. 7.14). Here the reflectance could be calculated by the FRESNEL formula (see, e.g., Born and Wolf 1975) for a certain angle of incidence θ_{in}.

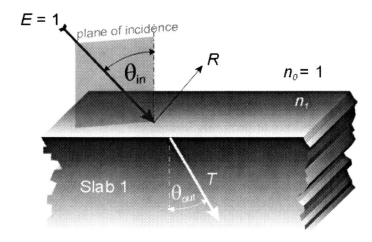

Fig. 7.14. Transmittance and reflectance at an optical boundary for non-perpendicular incidence ($\theta_{in} \neq 0°$).

The components of the direction of polarization parallel ($\parallel$) to or perpendicular ($\perp$) toward the angle of incidence are to be calculated separately from each other. The normalized reflections ($R_\parallel$ and $R_\perp$) are given as follows (neglecting the imaginary part of a complex refractive index, - see below):

$$R_\parallel = \frac{\tan^2(\theta_{in} - \theta_{out})}{\tan^2(\theta_{in} + \theta_{out})} \qquad R_\perp = \frac{\sin^2(\theta_{in} - \theta_{out})}{\sin^2(\theta_{in} + \theta_{out})} \tag{31}$$

$$T_\parallel = 1 - R_\parallel \qquad T_\perp = 1 - R_\perp \tag{32}$$

angle of refraction: $\theta_{out} = \arcsin\left(\dfrac{n_0}{n_1} \sin \theta_{in}\right)$ (33)

Complex refractive indices: For certain materials and wavelengths, the imaginary part k of the complex refractive index $\hat{n}=n-jk$ should not be neglected (in our case only at Silicon for $\lambda < 400$ nm). This could be done by an equation from Azzan and Bashara:

for perpendicular incidence ($R_0 = r_\parallel r_\parallel^* = r_\perp r_\perp^*$):

$$R_0 = \frac{(n-1)^2 + k^2}{(n+1)^2 + k^2}$$ (34)

for non-perpendicular incidences:

$$R_\parallel = r_\parallel r_\parallel^* = \frac{\mu + |\varepsilon|^2 x^2 - x(\mu + y^2)\sqrt{2(\mu+\kappa)}}{\mu + |\varepsilon|^2 x^2 + x(\mu + y^2)\sqrt{2(\mu+\kappa)}}$$ (35)

$$R_\perp = r_\perp r_\perp^* = \frac{\mu + x^2 - x\sqrt{2(\mu+\kappa)}}{\mu + x^2 + x\sqrt{2(\mu+\kappa)}}$$ (36)

$\mu^2 = (\epsilon - y^2)(\epsilon^* - y^2) = |\epsilon|^2 - 2\epsilon_1 y^2 + y^4$

$\epsilon_1 = n^2 - k^2$

$\epsilon_2 = 2 n k$

$|\epsilon|^2 = \epsilon \epsilon^* = \epsilon_1^2 + \epsilon_2^2 = (n^2 + k^2)^2$

$\kappa = \epsilon_1 - y^2$

$x = \cos(\theta_{in})$

$y = \sin(\theta_{in})$

The abbreviations κ, a, b and μ^2 are merely mathematical variables without any particular physical background.

7.1.3.2 Optical Transmittance of a Plane Slab

Incident insolation has to pass through two optical boundaries layers and the attenuation by absorption inside the material. Reflection at the lower boundary is not lost completely, but bounced up and forward through the slab at decreasing intensity (see Fig. 7.14).

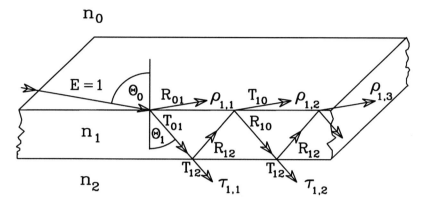

Fig. 7.15. Transmittance through a plane optical slab.

In order to establish a common nomenclature, a layer is marked by an index "k," the upper medium by the index "$k-1$" and the lower medium by "$k+1$". The angle of incidence at the transition from medium "$k-1$" to the medium "k" is marked by θ_k, the angle of refraction by θ_{k+1}. The optical transitions are denoted by the indices of media in the order of the radiation passing through them $k, k\pm 1$ (e.g. T_{01}). The transmitted parts i of the irradiance on slab k are marked as $\tau_{k,i}$, the reflected ones as $\rho_{k,i}$. A distinction between the planes of polarization is not made any more to keep the formulas simple. So, for the generally used variables R and T, the specific components $R_\parallel$, $R_\perp$ and $T_\parallel$, $T_\perp$ are to be inserted. The incident irradiance is also normalized to $E=1$. The absorptive attenuation of a ray after a passage through the slab 1 for example, is determined by the coefficient of absorption $\alpha_1(\lambda)$ of the material, its thickness d_1 and the incidence angle θ_1 on the considered slab 1:

$$\frac{\Delta E_1}{E_1} = \exp\left(-\alpha_1 \frac{d_1}{\cos \theta_1}\right) \tag{37}$$

If the imaginary part of the complex refractive index $\hat{n} = n - j\,k$ is known, α could also be computed from $\alpha(\lambda) = 4\pi k \lambda^{-1}$.

Therefore reflection loss R_{01} occurs at the incident boundary surface of slab 1 and R_{12} on its lower surface, while the remaining transmitted part consists of:

$$\tau_{l,1} = T_{01} T_{12} \exp\left(-\alpha_l \frac{d_1}{\cos\theta_l}\right) \tag{38}$$

The internal reflection R_{12} is computed the same as R_{0l} using (31), as well, the new angle of incidence θ_2 for the lower layer 2 according to (32). The internal reflection at the boundary 1-2 passes through layer 1 again, and is attenuated accordingly. At the boundary 0-1 this ray is refracted once more, while a component T_{10} passes into medium 0. The reflected component R_{10} reaches boundary 1-2 under attenuation, where another part T_{12} penetrates layer 2.

$$\tau_{l,2} = T_{01} R_{12} R_{10} T_{12} \exp\left(\frac{-3\alpha_l d_1}{\cos\theta_l}\right) \tag{39}$$

in general:
Summing up all transmitted fractions of the layer 1:

$$\tau_l = T_{01} T_{12} \exp\left(\frac{-\alpha_l d_1}{\cos\theta_l}\right) \sum_{i=1}^{\infty} \left[R_{12} R_{10} \exp\left(\frac{-2\alpha_l d_1}{\cos\theta_l}\right)\right]^{i-1} \tag{40}$$

$$\tau_{l,i} = T_{01} T_{12} R_{12}^{i-1} R_{10}^{i-1} \exp\left(\frac{-(2i-1)\alpha_l d_1}{\cos\theta_l}\right) \tag{41}$$

This infinite series is a geometrical one and can be summarized as follows:

$$\tau_l = \frac{T_{01} T_{12} \exp\left(\dfrac{-\alpha_l d_1}{\cos\theta_l}\right)}{1 - R_{12} R_{10} \exp\left(\dfrac{-2\alpha_l d_1}{\cos\theta_l}\right)} \tag{42}$$

Because radiation is being absorbed in the slab, the reflectivity ρ of a slab has to be computed explicitly, because $\rho \neq 1 - \tau$. The reflected components $\rho_{l,i}$ of a slab 1 are to be calculated as follows:

$$\rho_{l,1} = R_{01} \tag{43}$$

$$\rho_{1,2} = T_{01} R_{12} T_{10} \exp\left(\frac{-2\alpha_1 d_1}{\cos\theta_1}\right) \tag{44}$$

$$\rho_{1,3} = T_{01} R_{12}^2 R_{10} T_{10} \exp\left(\frac{-4\alpha_1 d_1}{\cos\theta_1}\right) \tag{45}$$

$$\rho_{1,i>1} = T_{01} R_{12}^{i-1} R_{10}^{i-2} T_{10} \left(\exp\left(\frac{-2\alpha_1 d_1}{\cos\theta_1}\right)\right)^{i-1} \tag{46}$$

The sum of all reflected components ρ_1 of the layer 1:

$$\rho_1 = R_{01} + T_{01} R_{12} T_{10} \exp\left(\frac{-2\alpha_1 d_1}{\cos\theta_1}\right) \sum_{m=1}^{\infty} \left[R_{10} R_{12} \exp\left(\frac{-2\alpha_1 d_1}{\cos\theta_1}\right)\right]^{m-1} \tag{47}$$

This infinite series is again a geometrical series and can be summarized:

$$\rho_1 = R_{01} + \frac{T_{01} R_{12} T_{10} \exp\left(\frac{-2\alpha_1 d_1}{\cos\theta_1}\right)}{1 - R_{10} R_{12} \exp\left(\frac{-2\alpha_1 d_1}{\cos\theta_1}\right)} \tag{48}$$

7.1.3.3 Internal Transmission and Reflection

Knowledge of the internal transmission $\bar{\tau}$ is necessary to determine the transmission of multiple layer systems, for example $\bar{\tau}_1$: the transmittance of slab 1, when it is illuminated from reflections coming out of slab 2. To distinguish internal transmittance and internal reflectance from the external ones, a bar over the variable is used.

$$\bar{\tau}_1 = T_{10} \exp\left(\frac{-\alpha_1 d_1}{\cos\theta_1}\right) \sum_{i=1}^{\infty} \left[R_{10} R_{12} \exp\left(\frac{-2\alpha_1 d_1}{\cos\theta_1}\right)\right]^{i-1} \tag{49}$$

$$\bar{\tau}_I = \frac{T_{10} \exp\left(\dfrac{-\alpha_I d_1}{\cos \theta_I}\right)}{1 - R_{10} R_{12} \exp\left(\dfrac{-2 \alpha_I d_1}{\cos \theta_I}\right)} \tag{50}$$

At the boundary between slab 2 and slab 1, T_{2I} is neglected because it has been already accounted for by the net reflectivity of the lower slabs. The internal reflectivity $\bar{\rho}_I$ is set up the same way:

$$\bar{\rho}_I = R_{10} T_{12} \exp\left(\frac{-2 \alpha_I d_1}{\cos \theta_I}\right) \sum_{i=1}^{\infty} \left[R_{12} R_{10} \exp\left(\frac{-2 \alpha_I d_1}{\cos \theta_I}\right) \right]^{i-1} \tag{51}$$

$$\bar{\rho}_I = \frac{R_{10} T_{12} \exp\left(\dfrac{-2 \alpha_I d_1}{\cos \theta_I}\right)}{1 - R_{12} R_{10} \exp\left(\dfrac{-2 \alpha_I d_1}{\cos \theta_I}\right)} \tag{52}$$

The transition from slab 2 to slab 1 (T_{2I}) is also neglected.

7.1.3.4 Transmittance Through Two Slabs

Now an optical system consisting of slab 1 and slab 2 will be examined. In addition to the internal reflections inside each slab there are also reflections over two slabs (between the upper boundary of the other slab and the lower boundary of the lower slab) to be considered. In order to keep the picture simple, the rays and their infinite series are summed up as slab transmittances τ and slab reflectances ρ. This is done in Fig. 7.16 and is marked by bold arrows. The following fraction outlines the most direct way into slab 3:

$$\tau_{12,1} = \frac{\tau_1 \tau_2}{T_{12}} \tag{53}$$

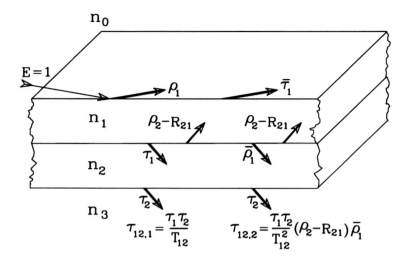

Fig. 7.16. Transmittance of an optical system consisting of two optical slabs.

It should be mentioned that for the combination of the slab transmittances τ_1 and τ_2 the reflection at the boundary 1-2 has been taken into account in τ_1 as well as in τ_2. So T_{12} has to be compensated once by τ_{12}. Accordingly this has to be done for further combinations of slabs τ_{k+1}.

The reflected part $\tau_1 \cdot (\rho_2 - R_{21}) \cdot T_{12}^{-1}$ from inside layer 2 enters slab 1 and is reflected at the boundary 1-0 back into slab 1 by losing $\bar{\tau}_1$. From slab 1 the fraction $\bar{\rho}_1$ reaches slab 2:

$$\bar{\rho}_1 = \frac{R_{10} T_{12} \exp\left(\frac{-2\alpha_1 d_1}{\cos \theta_1}\right)}{1 - R_{12} R_{10} \exp\left(\frac{-2\alpha_1 d_1}{\cos \theta_1}\right)} \tag{54}$$

The fraction $\bar{\rho}_1$ can be treated in the same way as the direct incoming part and therefore has the same attenuation $\tau_2 \cdot T_{12}^{-1}$ as that in slab 2 when it enters slab 3. This results in:

$$\tau_{12,2} = \frac{\tau_1 (\rho_2 - R_{12}) \bar{\rho}_1 \tau_2}{T_{12}^2} \tag{55}$$

Again a fraction $\rho_2 \cdot R_{12}$ is reflected from slab 2 into slab 1. Summarized, the total transmittance is:

$$\tau_{12} = \frac{\tau_1 \tau_2}{T_{12}} \sum_{i=1}^{\infty} \left[\frac{(\rho_2 - R_{12}) \bar{\rho}_1}{T_{12}} \right]^{i-1} \tag{56}$$

$$\tau_{12} = \frac{\tau_1 \tau_2}{T_{12} - (\rho_2 - R_{12}) \bar{\rho}_1} \tag{57}$$

The total reflectance ρ_{12} of the slab system can be derived accordingly:

$$\rho_{12,1} = \rho_1 \tag{58}$$

$$\rho_{12,2} = \frac{\tau_1 (\rho_2 - R_{12}) \bar{\tau}_1}{T_{12}} \tag{59}$$

$$\rho_{12,3} = \frac{\tau_1 (\rho_2 - R_{12})^2 \bar{\rho}_1 \bar{\tau}_1}{T_{12}^2} \tag{60}$$

$$\rho_{12} = \rho_1 + \frac{\tau_1 \bar{\tau}_1 (\rho_2 - R_{12})}{T_{12}} \sum_{i=1}^{\infty} \left[\frac{(\rho_2 - R_{12}) \bar{\rho}_1}{T_{12}} \right]^{i-1} \tag{61}$$

$$\rho_{12} = \rho_1 + \frac{\tau_1 \bar{\tau}_1 (\rho_2 - R_{12})}{T_{12} - (\rho_2 - R_{12}) \bar{\rho}_1} \tag{62}$$

The inner reflectivity $\bar{\rho}_{21}$ of the slab system is:

$$\bar{\rho}_{21} = \bar{\rho}_2 + \frac{\bar{\tau}_2 \bar{\rho}_1 \tau_2}{T_{12}} \sum_{i=1}^{\infty} \left[\frac{(\rho_2 - R_{12}) \bar{\rho}_1}{T_{12}} \right]^{i-1} \tag{63}$$

$$\bar{\rho}_{21} = \bar{\rho}_2 + \frac{\bar{\tau}_2 \tau_2 \bar{\rho}_1}{T_{12} - \bar{\rho}_1 (\rho_2 - R_{12})} \tag{64}$$

with (50):

$$\bar{\tau}_2 = \frac{T_{21} \exp\left(\dfrac{-\alpha_2 d_2}{\cos \theta_2} \right)}{1 - R_{21} R_{23} \exp\left(\dfrac{-2 \alpha_2 d_2}{\cos \theta_2} \right)} \tag{65}$$

7.1.3.5 Transmittance Through Three Slabs

The module encapsulation is now considered as three slabs of different optical properties (see Fig. 7.17). The two upper slabs 1 and 2 are now considered as a system described by its common transmittance and its common reflectance, as calculated already in the upper chapter. So interactions such as transmittances and internal reflections only need to be considered only between slab system 12 and the new slab 3.

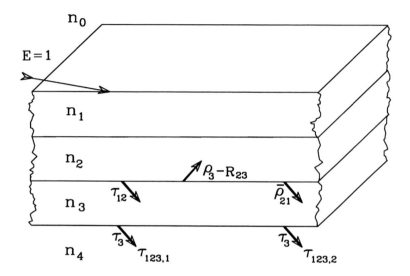

Fig. 7.17. Transmittance of an optical system consisting of three optical slabs.

After the split up of the ray at the module surface (i.e., slab 1) into a reflected and a transmitted part, the latter joins with the fractions being reflected at the inner boundaries at absorptive attenuation toward the boundary layer of slab 2. Accordingly, this is the same for the transmission of slab 2 and 3. Therefore the two upper slabs 1 and 2 are now considered as a system described by its common transmittance and its common reflectance. For the fraction entering slab 4 the following results occur:

$$\tau_{123,1} = \frac{\tau_{12}\,\tau_3}{T_{23}} \tag{66}$$

At the boundary 2-3 internal reflections from slab 3 pass to the border of slab system 21. (The reflection R_{23} has been accounted for already in τ_2 and is therefore subtracted.) A fraction passes through slab system 21, but another fraction will be reflected again and reaches slab 4 after passing boundary 2-3 and slab 3:

$$\tau_{123,2} = \tau_{12} \cdot \frac{(\rho_3 - R_{23})}{T_{23}} \cdot \frac{\bar{\rho}_{21} \tau_3}{T_{23}} \tag{67}$$

The sum of all fractions reaching medium 4 is:

$$\tau_{123} = \frac{\tau_{12} \tau_3}{T_{23}} \sum_{i=1}^{\infty} \left[\frac{(\rho_3 - R_{23}) \bar{\rho}_{21}}{T_{23}} \right]^{i-1} \tag{68}$$

$$\tau_{123} = \frac{\tau_{12} \tau_3}{T_{23} - (\rho_3 - R_{23}) \bar{\rho}_{21}} \tag{69}$$

7.1.3.6 Optical Transmittance Through m Slabs

All upper slabs 1 to $(m-1)$ are now considered as a system described by its common transmittance and its common reflectance, marked by the index $1...m$ (instead of : 123...to m). For the first fraction entering slab m+1 the following result occurs:

$$\tau_{1...m,1} = \frac{\tau_{1...m-1} \cdot \tau_m}{T_{(m-1)m}} \tag{70}$$

At the boundary $(m-1)$ m internal reflections from slab m pass to the border of the slab system $(m-1)..1$, The reflection $R_{(m-1)m}$ has been accounted for already in $\tau_{(m-1)}$ and is therefore subtracted. One fraction passes through slab system $(m-1)...1$, but another fraction will be reflected again and reaches slab $(m+1)$ after passing the boundary $(m-1)$ m and slab m:

$$\tau_{1...m,2} = \tau_{1...(m-1)} \cdot \frac{(\rho_m - R_{(m-1)m})}{T_{(m-1)m}} \cdot \frac{\bar{\rho}_{(m-1)...1} \tau_m}{T_{(m-1)m}} \tag{71}$$

The sum of all fractions reaching medium $(m+1)$ is:

$$\tau_{1...m} = \frac{\tau_{1...m} \cdot \tau_m}{T_{(m-1)m}} \sum_{i=1}^{\infty} \left[\frac{(\rho_m - R_{(m-1)m}) \cdot \bar{\rho}_{(m-1)...1}}{T_{(m-1)m}} \right]^{i-1} \tag{72}$$

$$\tau_{1...m} = \frac{\tau_{1...(m-1)} \cdot \tau_m}{T_{(m-1)m} - (\rho_m - R_{m(m-1)}) \cdot \bar{\rho}_{(m-1)...1}} \tag{73}$$

This $\tau_{1..m}$ consists of the components $\tau_{1..m\parallel}$ and $\tau_{1..m\perp}$ which are to be multiplied by the components of the according directions of polarization. The inner reflectivities are:

$$\bar{\rho}_{m...1} = \bar{\rho}_m + \frac{\bar{\tau}_m \cdot \tau_m \cdot \bar{\rho}_{(m-1)...1}}{T_{m(m-1)} - \bar{\rho}_{(m-1)...m} \cdot (\rho_m - R_{(m-1)m})} \tag{74}$$

Another solution for the transmittance of multi-layer systems is a complex matrix procedure proposed by Klein and Furtak (1988), respectively Shabana and Namour (1990). To understand the optical process in more detail, the model presented above is more appropriate.

7.1.3.7 Simulation results

Actual transmittance as a function of the incidence angle for an optical system consisting of three slabs using the model described is plotted in Fig. 7.18. The refractive indices and the absorption coefficients that were used are given in Table 7.4 for a wavelength of λ = 800 nm. Thicknesses used are 2 mm for glass, 0.5 mm for EVA and 0.05 mm for TiO_2.

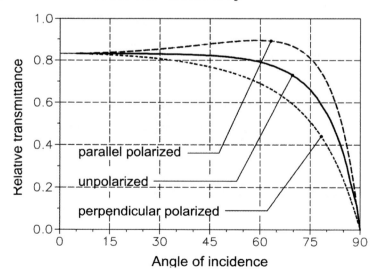

Fig. 7.18. Transmittance of three optical slabs as a function of the incidence angle.

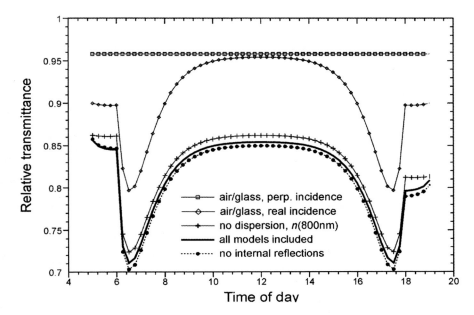

Fig. 7.19. Transmittance through a PV module during a clear day using different models (no absorption).

Figure 7.18 shows the transmittance of the same optical system over the time of day (21st of March) for an anisotropic sky hemisphere – applying the model described in DIN 5034 – using the actual spectra (interpolated from CIE standard spectra) for clear sky conditions. Refractive indices are given in Table 7.4, absorption in the upper three layers was set to zero. The simulation has been carried out for a location 30° South of the Equator at an elevation angle of the optical system of 30° from horizontal.

A remarkable effect is that optical performance drops significantly after sunrise (at 6 a.m.) and before sunset (at 6 p.m.). These minima can be explained by high reflection losses for the direct component due to flat incidence angles. Before sunrise and after sunset just a diffuse component exists which suffers from relatively small reflection loss. At noon transmittance is maximal. Additionally, the effects of some simplifications to the model have been evaluated: Relatively small deviations from the original track could be observed if the internal reflections are neglected; this measure leads to an underestimation of the yield in the vicinity of 1%. Neglecting optical dispersion results in an overestimation of the yield of about 1.5%. Disregard of lower slabs and the use of perpendicular incidence only causes vast overestimations of transmittance and therefore should not be applied for modeling.

Table 7.4. Refractive indices and absorption coefficients of the materials used

Materials (along path of irradiance into module)	Wavelength λ in nm	Refractive index n	Absorption coefficient α in m^{-1}	Reference
Glass (Optiwhite)	400	1.54	1.22	Flachglas 1989
	550	1.52	4.56	
	600	1.52	1.96	
	800	1.52	4.38	
	1000	1.51	5.31	
EVA (Elvax 150, lamination for 20 min at 149°C)	400	1.49	323.20	Gueris 1991
	600	1.47	26.15	
	800	1.45	23.65	
	1000	1.44	25.20	
	1200	1.43	25.75	
SiN$_x$ (hydrogenized silicon nitride SiN$_x$:H)	300	2.10–3.10 [a]	[a]	Lelièvre et al. 2005
	400	2.23–3.42 [a]	[a]	
	500	2.00–3.30 [a]	[a]	
	600	1.98–3.06 [a]	[a]	
	700	1.95–2.92 [a]	[a]	
	850	1.93–2.87 [a]	[a]	
TiO$_2$ (deposited at 300°C)	400	2.73	due to the very thin layer of TiO$_2$, absorption is negligible	Jellison et al. 1985
	550	2.43		
	600	2.39		
	800	2.30		
	1000	2.27		
ZnO	589	2.0036		CRC 2005
	750	1.9662		
	1000	1.9435		

Materials (along path of irradiance into module)	Wavelength λ in nm	Refractive index n	Absorption coefficient α in m^{-1}	Reference
Silicon*	354	5.61		CRC 1994
	400	5.57	10^7	Palik 1985
	496	4.32	10^4	CRC 1994
	600	3.95		Palik 1985
	729	3.75	$2 \cdot 10^3$	CRC 1994
	800	3.69		Palik 1985
	1000	3.57	10^2	

[a] the refractive index can be adjusted during processing via the precursor gas ratio NH_3/SiH_4, the absorption coefficients also depend strongly on that ratio: the lower the ratio, the higher is the refractive index and the higher is the absorption coefficient.

* the complex part k of the refractive index n of Silicon is (Palik 1985): $k(\lambda=350nm)=2.99$; $k(\lambda=400nm)=0.39$; $k(\lambda=500nm)=0.07$; $k(\lambda=600\ nm)=0.03$; $k(\lambda=800\ nm)=0.01$

7.2 Model to Determine the Actual Cell Temperature

The actual electrical output power of a solar module with silicon solar cells depends not only on the solar radiation being absorbed and transformed, but also on the actual operating temperature of the cells. For an increase of temperature power output and yield decreases by ca. -0.5%/K (see Table 5). Therefore it is important to determine the operating cell temperature over the day. The thermal model presented starts with the conduction of a given heat flow from the cells to the module surface and then its distribution from the module surface to the environment by convection and radiation. The combination of optical, electric and thermal modeling pays account to the following processes and parameters: Irradiance, reflection, transmittance, absorption, photovoltaic energy conversion, balance of heat flows, thermal composition of module, ambient temperature, wind speed and equivalent sky temperature to determine radiation-exchange. To simulate the worst-case conditions, a focus of the modeling is on natural convection. A scheme of the optical, electrical and thermal energy flows is presented in Fig. 7.20.

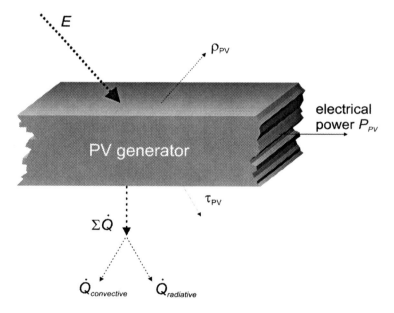

Fig. 7.20. Scheme of the incoming and outgoing energy flows of a PV module (optical energy flows are plotted as doted lines, thermal as broken lines and electrical as full lines, the width represents the typical value of the flows).

It is assumed that the cells are opaque and all absorbed radiation minus the photovoltaic conversion is contributing to heat flow generation. All absorption happens in the solar cell, absorption and heat flow generation in the encapsulation layers are neglected. This is valid, because the materials used are optimized for high transmittance, therefore absorption is very low (e.g., absorption for Optiwhite® glass of 2 mm of thickness: < 0.5% for λ = 400–1,200 nm, at EVA (Elvax® 150) < 1.5% for λ = 450–1,100 nm).

7.2.1 Heat Flow Input

Total energy flow onto the module surface consists of the following components:

1a) Temperature radiation exchange of the module surface with the sky (consists of sky radiation and heat radiation of the atmosphere).

1b) Temperature radiation exchange of the module surface with the surface of the ground.

2) Heat transfer from the air to the module (module has at least ambient temperature during the day).

3a) Direct terrestrial solar irradiance (0.35 µm < λ < 2 µm; unpolarized, $V = 0\ \%$)

3b) Diffuse terrestrial solar irradiance (0.35 µm < λ < 2 µm; polarized, $V \leq 70\ \%$)

3c) Albedo (reflected solar radiation from the earth's surface onto the module; depends very much on the properties of the surrounding earth's surface and is difficult to calculate exactly)

That total energy flow is partly reflected by the module encapsulation and the solar cell (see chapter 7.1.3) or converted into electricity and distributed to the terminal. The remaining energy flow absorbed in the cell is the heat flow input.

7.2.1.1 Heat Flow Input by Sky and Ground Radiation

While the temperature of the ground and the sky are in general lower than the temperature of the module during daytime, the energy input to the module by radiation exchange is negative, and therefore it will be treated in chapter 7.2.3.

7.2.1.2 Heat Flow Input by Ambient Temperature

Due to the relatively low specific heat capacity ($C_p = 7.3$–11.3 kJ m^{-2} K^{-1}) of the module and the relatively good heat transfer coefficient, the module reaches ambient temperature quickly (exception: see chapter 10.4.3. to 10.4.6.). Measurements of temperature and irradiance show some correlation between irradiance and ambient temperature. We use a simplified approach presented by Abid 1987 showing a cosine temperature course with a phase shift of t-t_0 hours against the course of the temperature during the day. While Abid 1987 suggests a maximum of ambient temperature at $t_0 = 18$ hours for hot arid climatic zones, more recent examinations by Farkas 1992 suggest a $t_0 = 14$ hours. After comparison with measured data the suggestion from Farkas 1992 was implemented, which did not show the minimum temperature at sunrise (as measured), but its maximum fits well to the measurements.

$$T_A = \bar{T}_A + \left(T_{A,\max} - \bar{T}_A\right)\cos\left(\frac{\pi}{12}(t-t_0)\right) \tag{75}$$

T_A ambient temperature in K

$T_{A,max}$ maximum of ambient temperature during the day in K

$\bar{T}_A$ average ambient temperature in K

t time of day in hrs

t_0 time of the maximum temperature in hrs

According to Rauschenbach 1980 extremely high ambient temperatures (> 50°C) in hot arid zones occur at very low wind speeds (< 1 m s^{-1}). Prerequisite for such temperatures are local accumulations of hot air at high elevation angles of the sun (low AM). Rauschenbach 1980 mentions that with the help of albedo ground reflections temporarily air temperatures of up to 70°C are possible. In non-arid climatic zones the course of temperature depends very much on weather and season, so for such zones with changing conditions the use of measured temperature data sets is recommendable.

7.2.1.3 Heat Flow Input by Irradiance

Incoming irradiance consists of direct irradiance and of the diffuse irradiance, being scattered in the Earth's atmosphere. This incoming radiation serves as input for the optical model described in chapter 7.1.3. Components of irradiance, which reach the cell but could not be converted into electricity through the photovoltaic process, are main participators to the heat flow input. If ground reflection (Albedo) is not considered, the following equation is true:

$$\dot{Q} = P_{in} \cdot (1 - \eta_{PV}) = A \cdot E_{cell} \cdot (1 - \eta_{PV}) \tag{76}$$

A	front surface area in m^2
E_{cell}	cell reaching irradiance in W m^{-2}
P_{in}	irradiance power input to the solar cells in W
Q	total heat flow in the module in W
η_{PV}	photovoltaic conversion efficiency of the solar cells

$$E_{cell} = 0.5 \cdot E_{dir}(\tau_{123,\parallel} + \tau_{123,\perp}) + \int_0^{\frac{\pi}{2}} \int_0^{2\pi} E_{dif,\parallel}(\alpha_P,\gamma_P) \cdot \tau_{123,\parallel} + E_{dif,\perp}(\alpha_P,\gamma_P) \cdot \tau_{123,\perp} d\alpha_P d\gamma_P \tag{77}$$

E_{dir}	direct irradiance onto the module surface in W m^{-2}
$E_{dif}(P)$	diffuse radiance from a point P at the sky hemisphere onto the module surface in W m^{-2}
α_P	azimuth angle of point P in °
γ_P	elevation angle of point P in °
τ_{123}	optical transmittance of the full encapsulation
$\parallel$	polarization parallel to the plane of incidence
$\perp$	polarization perpendicular to the plane of incidence

The part being dissipated electrically not only depends on the possible conversion efficiency, but also on the adaption of the load resistance to the I-V characteristics of the solar cell. For simplified reasons we assume that the maximum available electrical power is always taken from the terminals. A worst-case scenario would suggest the use of short-circuit or open-circuit conditions, due to the elevated heat generation, but these conditions are atypical for the common use of solar generators.

7.2.2 Heat Transfer Inside a Module

The absorbed component of irradiance that is not able to be electrically dissipated generates a heat flow. That heat flow has to get to the surface of the module (stationary heat conduction) or increases the module's temperature (non-stationary heat flow).

7.2.2.1 Dimensional Layout of the Thermal Model

In relation to the length and the width of a module, the path for heat transfer perpendicular through the layers is much shorter. Thus a one dimensional modeling of the layers (assuming an infinite surface) could be applied. This simplification is even more suitable while the area of the modules was increasing within the last years (see Table 7.5). Due to cost reduction efforts, framing and support structure are also minimized, which increases accuracy of the one-dimensional assumption.

Table 7.5. Development of size and power of PV modules

Year	Area in m^2 (sizes in m)	Power in W$_p$	Type; Manufacturer	Remarks
1979	0.17 (0.47 x 0.37)	10.9	BPX 47A; Valvo	round cells, sc-Si
1979	0.27 (0.58 x 0.47)	18.3	BPX 47B/20; Valvo	round cells, sc-Si
1983	0.50 (1.08 x 0.46)	50	PQ 40/50; AEG	now: Schott/ASE, mc-Si
1985	1.50 (1.47 x 1.02)	130	SM 144; Siemens	round cells, sc-Si
1991	1.11 (1.12 x 0.99)	120	MSX-120; Solarex	series prod., mc-Si
1991	1.84 (1.10 x 1.68)	188	PS 184 T; Nukem	MIS-I
1992	0.94 (1.10 x 0.86)	96	PS 94 MP 96; Nukem	series prod., MIS-I
1992	0.82 (1.27 x 0.64)	100	M 100 L; Siemens	series prod., sc-Si
1994	2.42 (1.89 x 1.28)	285	ASE-300-DG/50; ASE	series prod., EFG-Si
2003	4.5	500	P 500; Solon AG	Prototype, mc-Si
2005	12.5	1,500	Ertl Glass, Austria	Prototype

7.2.2.2 Stationary Heat Flow in the Module

Heat flow by conduction inside the module from the cell through three layers to the module font-surface respectively to its back surface is given by:

$$\dot{Q}_F = A\left(T_C - T_F\right) \sum_{i=1}^{3} \frac{k_i}{d_i} = \frac{T_C - T_F}{R_{k,F}} \qquad (78)$$

$$\dot{Q}_B = A\left(T_C - T_B\right) \sum_{i=4}^{6} \frac{k_i}{d_i} = \frac{T_C - T_B}{R_{k,B}} \qquad (79)$$

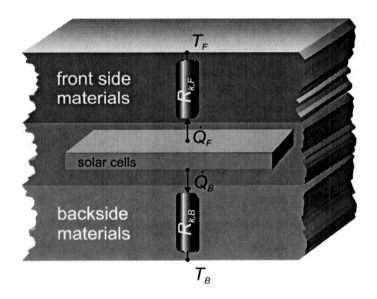

Fig. 7.21. Thermal equivalent electrical circuit for heat conduction in a PV module.

A	front- and backside surface area of the module in m^2
d_i	thickness of the layer i in m
$\dot{Q}_F$	heat flow from the cells to the front surface in W
$\dot{Q}_B$	heat flow from the cells to the back surface in W
$R_{k,F}$	thermal resistance from the cells to the front surface in K W^{-1}
$R_{k,B}$	thermal resistance form the cells to the backside in K W^{-1}

T_C cell temperature in K

T_F temperature of the front surface in K

T_B temperature of the back surface of the module in K

k_i thermal conductivity of the layer i in W m^{-1} K^{-1}

Cell temperature T_C could be computed by knowing (or measuring) the front- and backside surface temperature (T_F and T_R) and the heat flow $\dot{Q}$:

$$T_C = \frac{\dot{Q} R_{k,F} R_{k,B} + T_F R_{k,B} + T_B R_{k,F}}{R_{k,F} + R_{k,B}} \qquad (80)$$

7.2.2.3 Non-Steady-State Heat Flow in the Module

The materials used are not just showing specific thermal conductivities, but also specific heat capacities c_p. For the heat flow $\dot{Q}_{sto}$, which is being stored, the following relationship is valid:

$$\dot{Q}_{sto} = C_p \frac{dT}{dt} = c_p \rho d_i A \frac{dT}{dt} \qquad (81)$$

A surface area in m^2

c_p specific heat capacity of the material in J kg^{-1} K^{-1}

C_p total heat capacity of them material in J K^{-1}

d_i thickness of the layer i in m

$\dot{Q}_{sto}$ heat flow into the storage in W

ρ density of the material in kg m^{-3}

Therefore cell temperature T_C can be calculated as follows:

$$T_C = \frac{(\dot{Q} - \dot{Q}_{sto}) R_{k,F} R_{k,B} + T_F R_{k,B} + T_R R_{k,F}}{R_{k,F} + R_{\lambda,B}} \qquad (82)$$

7.2.3 Heat Dissipation

Heat dissipation of the heated module consists of the following components:

1.) heat transfer by thermal radiation exchange with sky and ground,
2.) heat transfer by convection,
3.) heat transfer by thermal conduction towards a heat capacity.

Heat transfer by thermal conduction from the module to the support structure is neglected. While this normally occurs just at the module framing, it depends very much on the type of mounting, thus a one-dimensional model is used (see chapter 7.2.2.1).

7.2.3.1 Thermal Radiation Exchange with Sky and Ground

Thermal radiation exchange between the module and the sky always occurs when the module surface temperature is higher than the sky temperature and the module surface shows an $\varepsilon > 0$. While glass is almost opaque for far infrared radiation according to Scholze 1988 (p. 337), thermal radiation exchange by the cell can be neglected for temperatures lower than 230°C. So thermal radiation just occurs from the module surface, where the Stefan-Boltzmann's law of thermal radiation can be applied:

$$\dot{Q}_{rad} = \sigma \varepsilon A T_M^4 \tag{83}$$

$\dot{Q}_{rad}$ thermal radiation heat flow in W

A area of radiating surface in m^2

T_M module surface temperature in K

σ Stefan-Boltzmann's constant ($\sigma = 5.67 \cdot 10^{-8}$ W m^{-2} K^{-4})

$\varepsilon = f(T_M)$ emissivity of the surface as a function of surface temperature (in IR)

The wavelength at which maximum emission of a black radiator ($\varepsilon = 1$) occurs is given by *Wien's Displacement Law*:

$$\lambda_{max} = \frac{2{,}887{,}800}{T} \text{ K nm} \tag{84}$$

λ_{max} wavelength at maximum of emission in nm

E.g., for a black body of $T=75\,°C$ the wavelength of maximal emission is ca. 8,300 nm, at a black body of $T=6,000$ K (sun's surface) is ca. 550 nm. Emissivities $\varepsilon\ (T_M)$ of typical construction materials of PV modules are given in Table A14 in the Appendix.

Heat flow by radiation exchange $\dot{Q}_{12}$ between two grey ($0 < \varepsilon < 1$) bodies with the temperatures T_1 resp. T_2 is given as:

$$\dot{Q}_{12} = \frac{\sigma\,\varepsilon_1\,\varepsilon_2 A_1\,\varphi_{12}}{1-(1-\varepsilon_1)(1-\varepsilon_2)\,\varphi_{12}\,\varphi_{21}}(T_1^4-T_2^4) \tag{85}$$

The so called view factor φ_{12} has to be considered separately for all possible combinations of radiation exchange of the front- and backside of the module, the sky and the ground (see Fig. 7.22.). For the view factor from the font to the sky (φ_{FS}) and from the back to the sky (φ_{BS}), as well as for the view factor from the front to the ground (φ_{FG}) and from the back to the ground (φ_{BG}), the following equations can be derived (γ_M is the elevation angle of the module):

$$\varphi_{FS} = 0.5\,(1 + \sin(90° - \gamma_M)) \tag{86}$$

$$\varphi_{FG} = 0.5\,(1 - \sin(90° - \gamma_M)) \tag{87}$$

$$\varphi_{BS} = 0.5\,(1 - \cos \gamma_M) \tag{88}$$

$$\varphi_{BG} = 0.5\,(1 + \cos \gamma_M) \tag{89}$$

For the reverse view factor φ_{21} (i.e., from the environment to the module) the following equation is valid:

$$\varphi_{21} = \frac{A_1}{A_2}\,\varphi_{12} \tag{90}$$

While the area of the environment A_2 is infinite and the reverse view factor is getting zero, equation (85) simplifies to:

$$\dot{Q}_{12} = \sigma\,\varepsilon_1\,\varepsilon_2 A_1\,\varphi_{12}(T_1^4-T_2^4) \tag{91}$$

Heat transfer by radiation consists of the following components:

$$\dot{Q}_{FS} = \sigma\,\varepsilon_F\,\varepsilon_S A_F\,\varphi_{FS}(T_F^4-T_S^4) \tag{92}$$

$$\dot{Q}_{FG} = \sigma\,\varepsilon_F\,\varepsilon_G A_F\,\varphi_{FG}(T_F^4-T_G^4) \tag{93}$$

$$\dot{Q}_{BS} = \sigma\,\varepsilon_B\,\varepsilon_S A_B\,\varphi_{BS}(T_B^4-T_S^4) \tag{94}$$

$$\dot{Q}_{BG} = \sigma \varepsilon_B \varepsilon_G A_B \varphi_{BG} (T_B^4 - T_G^4) \qquad (95)$$

The entire heat radiation exchange is the sum of these components:

$$\dot{Q}_{rad} = \dot{Q}_{FS} + \dot{Q}_{FG} + \dot{Q}_{BS} + \dot{Q}_{BG} \qquad (96)$$

Using (86)-(89) in (92)-(95) and then including that into (96), the result is:

$$\begin{aligned}\dot{Q}_{rad} = \tfrac{1}{2}\sigma[A_F \varepsilon_F (\ &\varepsilon_S(T_F^4 - T_S^4)(1+\sin(90°-\gamma_M))\\ &+ \varepsilon_G(T_F^4 - T_G^4)(1-\sin(90°-\gamma_M)))\\ + A_B \varepsilon_B (\ &\varepsilon_S(T_B^4 - T_S^4)(1-\cos\gamma_M)\\ &+ \varepsilon_G(T_B^4 - T_G^4)(1+\cos\gamma_M))]\end{aligned} \qquad (97)$$

For small elevation angles of the module (e.g., for locations close to the equator) $\dot{Q}_{FG}$ and $\dot{Q}_{BS}$ can be neglected, while $\varphi_{FS} = \varphi_{BG} = 1$, so the total heat radiation exchange simplifies to:

$$\dot{Q}_{rad}(\gamma_M \approx 0°) = \dot{Q}_{FS} + \dot{Q}_{BG} = \sigma\left(\varepsilon_F \varepsilon_S (T_F^4 - T_S^4) + \varepsilon_B \varepsilon_G (T_B^4 - T_G^4)\right) \qquad (98)$$

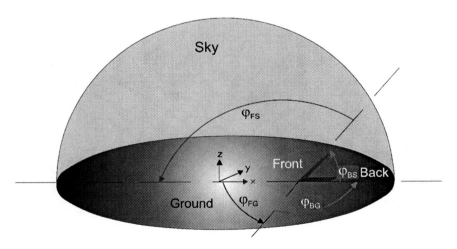

Fig. 7.22. Definition of view factors φ for the thermal radiation exchange between a terrestrial solar module and the sky respectively the ground.

7.2.3.2 Determination of Sky Temperature

The sky temperature T_H (equivalent temperature of the uncovered sky for $\varepsilon_s = 1$) is given by Swinbank 1963 as:

$$T_S = 0.0552 \cdot T_A^{1.5} \tag{99}$$

T_S sky temperature in K
T_A ambient temperature in K

or according to an even more simplified assumption by Whiller 1967:

$$T_S = T_A - 6\,\text{K} \tag{100}$$

A more extensive, but more accurate method (using also relative air humidity HR as parameter) is given by Abid 1987:

$$T_S = T_A \sqrt[4]{(5.7723 + 0.9555 + (0.6017)^7) \cdot 10^{-4} T_A^{1.1893} HR^{0.0665}} \tag{101}$$

A comparison of the different models for the calculation of sky temperature T_S is given in Fig. 7.23 as a function of ambient temperature T_A. In general, further calculations were using the model from Abid 1987 at $HR=40\%$.

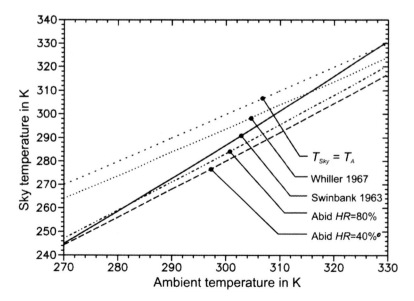

Fig. 7.23. Comparison of sky temperatures T_S as a function of ambient temperatures T_A computed by different models.

7.2.3.3 Heat Dissipation by Natural Convection

Heat transfer by natural convection occurs when the surface temperature of the module T_M is higher than the ambient temperature T_A and convection is possible at least at one module surface. Convective heat flows cannot be calculated exactly by mathematics and therefore have to be computed by iteration or by approximation equations such as shown here. First, some important characteristics which are relevant for convection:

The Prandtl number Pr is a characteristic which compares two molecular transport mechanisms: impulse transfer by friction and heat transfer by thermal conduction. The Prandtl number is a function of temperature and depends on the physical properties of the fluid. For gas molecules with the same number of atoms, the Prandtl number remains constant.

The Reynolds number Re determines the heat transfer at forced convection. Its value also settles the type of flow – whether it is turbulent or laminar. The type of flow is of decisive significance: At a laminar flow no mixture movement occurs and heat transfer is low. In zones of turbulent flow the thickness of the Prandtl's border layer depends on the Reynolds number. The "abrasive effect" of the turbulent flow core (flowing with a velocity w) tries to lower the thickness of the laminar border layer; viscosity v works in opposition to that, supported by the roughness of the wall (which is not directly included in the Reynolds number).

The Grashof number Gr provides the relation of thermal buoyancy force to inner "sluggishness." In case of free flow conditions only natural convection is accountable for the heat transfer. The temperature difference between the wall and the average temperature of the flowing media, which causes a change in volume and therefore a change in density at some parts of the media, is the driving force for the natural convection.

The Nußelt number Nu shows the relation of the actual heat flux density to pure heat conduction through a layer equivalent to the relevant length l at which flowing occurs.

Another substantial parameter for natural convection is, beside Nußelt number Nu and the Prandtl number Pr, the Rayleigh number Ra. If Ra exceeds a certain threshold, the so called "critical Rayleigh number" Ra_{cr}, the convective border layer is lifted off, which makes a case division necessary for the convective heat transfer. Ra is calculated as follows:

$$Ra = Pr \cdot Gr = Pr \frac{g \beta c_p \rho l_{ch}^3}{\eta k}(T_M - T_A) = Pr \frac{g \beta l_{ch}^3}{Nu}(T_M - T_A) \qquad (102)$$

Gr Grashof number

Nu Nußelt number

Pr Prandtl number
g ground acceleration ($g = 9.8067$ m s^{-2})
β thermal expansion coefficient of air in K^{-1}
η dynamic viscosity in Pa s (or N m^{-2} s)
l_{ch} length of a body relevant for the flow in m
T_A ambient temperature in K
T_M surface temperature of the module in K

Ra_{cr} can be computed by the following approximation:

$$Ra_{cr} = 10^x$$
$$\text{with:} \quad x = 8.9 - 0.00178 \cdot (90° - \gamma_M)^{1.82} \tag{103}$$

For the Prandtl number Pr:

$$Pr = \frac{\eta \, c_p}{k} \tag{104}$$

c_p specific heat capacity in N m kg^{-1} K^{-1}
η dynamic viscosity of air in Pa s
k heat conductivity of air in W m^{-2} K^{-1}
x auxiliary variable
γ_M elevation angle of module in °

Inserting (103) into (102) using the parameters of air relevant for our problem from Table A15, page 251 (with $l_{ch} = 1.3$ m, $\gamma_M = 45°$), for $Ra > Ra_{cr}$ the following equation has to be fulfilled: $T_M - T_A > 0.05$ K, which is always true during the day, so we can take boundary layer being lifted off for granted. For the upper surface of a tilted (plane) plate the Nußelt number is for $Ra > Ra_{cr}$ (according to Fujii et al. 1972):

$$Nu_F = 0.56 \sqrt[4]{Ra_{cr} \sin \gamma_M} + 0.13 \left(\sqrt[3]{Ra} - \sqrt[3]{Ra_{cr}} \right) \tag{105}$$

The convective heat transfer coefficient for natural convection (nc) at the front (F) surface is:

$$h_{nc,F} = \frac{Nu_F \, k}{l_{ch}} \tag{106}$$

In opposition to the front surface of the module a lifting-off of the boundary layer is not possible on the backside of the module, while the heated air is subject to buoyancy and can just escape along the rear surface of the module. For Nu at non-lifted-off boundary layer for $\gamma_M \leq 90°$ the equation by Merker 1987 is valid, if Ra is substituted by $Ra \sin\gamma_M$:

$$Nu_B = 0.56 \sqrt[4]{Ra \sin\gamma_M} \qquad (107)$$

so the convective heat transfer coefficient for the backside (B) is:

$$h_{nc,B} = \frac{Nu_B k}{l_{ch}} = 0.56 \frac{k}{l_{ch}} \sqrt[4]{Ra \sin\gamma_M} \qquad (108)$$

The main problem for further calculations is that the heat transfer coefficients are functions of the surface temperatures T_F and T_B. These temperatures can be computed only after the calculation of the heat transfer coefficient because it depends on the balance of heat fluxes. Therefore, an iteration procedure has to be applied. This also allows the use of the temperature of the boundary layer (average of ambient- and surface-temperature) instead of the ambient temperature T_A. Estimates for starting values for the iteration process have to be given. A good estimate is $T_F = T_B = T_A + 30$ K.

Relevant properties for air (Pr, k, c_p, ρ) at different temperatures are given in the Annex in Table A10. In a very simplified model, given by Funck 1985 (applied for the thermal layout gear boxes), heat transfer coefficients (for natural convection at small bodies) of horizontal and vertical surfaces are equalized. While the elevation angle of solar modules is between the extreme values of horizontal (PV systems at the equator) and vertical (PV facades), and further on the course of the function (elevation angle vs. heat transfer coefficient) is continuous, Funck's approximation can be used as an elevation-angle-independent guess for heat transfer by natural convection at small solar modules:

$$h_{nc} = \frac{12.1 \left(\frac{T_M - T_A}{T_M}\right)^{0.3}}{10\sqrt{l_{ch}}} \qquad (109)$$

h_{nc} heat transfer coefficient for natural convection in W m^{-2} K^{-1}

l_{ch} length of module over which convection take place in m

T_A ambient temperature in K

T_M module surface temperature in K

The length l_{ch} over which convection occurs does not have a significant effect on the heat transfer coefficient h_{nc}, while its length is influencing the result only by the tenth root of its reciprocal value. This means that for all lengths of PV modules available on the market for power applications (0.5 m < l_{ch} < 2 m), the variation of the heat transfer coefficient for natural convection is only ±5 %.

7.2.3.4 Heat Dissipation by Forced Convection

In opposition to natural convection, the so called buoyancy term $g\ \beta\ \Delta T$ in equation (102) is negligible and thus the characteristics which contain a part of the buoyancy term(Ra and Gr) cannot be developed. Therefore, at forced convection, only the characteristics Nu, Re, and Pr are of significance.

For the case of a flow parallel to the surface the characteristic length l_{ch} is equal to the length of module L if the direction of the wind is along its length. If the direction of the wind is parallel to the width of the module $l_{ch} = B$ (width of the module). The first case can occur only at a horizontal installation of the module ($\gamma_M = 0°$) or at redirected wind flow (e.g., at very large PV power plants or at hillside locations). The second case occurs, when the wind direction is perpendicular to the azimuthal orientation of the module (i.e., westerly and easterly winds). All other cases are to be considered as "flow oblique of perpendicular to the plate" (see below).

Depending on the flow speed, a laminar or turbulent boundary layer may come into existence. Due to higher heat transfer coefficients and Nußelt numbers, heat exchange is by a factor of two better at the turbulent case. Turbulence condition is $Re > 10^5$ and is true for wind speed according to:

$$w > \frac{10^5 v}{l_{ch}} \tag{110}$$

using $T_A = 293.15$ K, the turbulence condition is true for $w > 1.33$ m s^{-1}, which is equal to Beaufort 2. At a turbulent boundary layer the following equations are valid (see, e.g., Merker 1987):

$$Nu_{tur} = \frac{0.037\ Re^{0.8} Pr}{1 + 2.433\ Re^{-0.1} \left(\sqrt[3]{Pr^2} - 1\right)} \tag{111}$$

$$h_{tur} = \frac{k}{l_{ch}} Nu_{tur} = \frac{0.037\ Re^{0.8} Pr\ k}{l_{ch}\left(1 + 2.433\ Re^{-0.1} \left(\sqrt[3]{Pr^2} - 1\right)\right)} \tag{112}$$

For wind speeds lower than 1.33 m s^{-1} a laminar boundary layer can occur, if the flow has not been turbulent already before reaching the plate. For a pure laminar boundary layer the heat transfer coefficient is:

$$h_{lam} = \frac{k}{l_{ch}} Nu_{lam} = \frac{0.664\, k \sqrt{Re}\, \sqrt[3]{Pr}}{l_{ch}} \qquad (113)$$

While the flowing conditions, especially at the transition from laminar to tubular boundary layers, are often (e.g., caused by small obstacles on the surface or by blunt-edges plates) in parts laminar, in parts turbulent, VDI-Wärmeatlas 1991 recommends for $10^1 < Re < 10^7$ (i.e., $0.13 \cdot 10^{-3}$ m s^{-1} $< w <$ $1.33 \cdot 10^2$ m s^{-1}) to use an average Nußelt-number (respectively heat transfer coefficient):

$$\overline{Nu_{fc}} = \sqrt{Nu_{lam}^2 + Nu_{tur}^2} \qquad (114)$$

$$h_{fc} = \sqrt{\left(\frac{0.664\, k \sqrt{Re}\, \sqrt[3]{Pr}}{l_{ch}}\right)^2 + \left(\frac{0.037\, k\, Re^{0.8}\, Pr}{l_{ch}\left(1 + 2.433\, Re^{-0.1}\left(\sqrt[3]{Pr^2} - 1\right)\right)}\right)^2} \qquad (115)$$

7.2.3.5 Heat Transfer for Superposition of Natural and Forced Convection

For wind speeds $w \le 1$ m s^{-1} the heat flow by natural convection exceeds the one caused by forced confection. Because total calm happens only under exceptional conditions, heat transfer by forced convection could not be neglected most of the time. The stance in literature to treat both convection types separate could not be followed. A suggestion developed from experimental results is given by Churchill 1977: For the case of equal direction of free and forced convection the following equation is valid:

$$Nu = \sqrt[3]{Nu_{fc}^3 + Nu_{nc}^3} \qquad (116)$$

For the case of opposite directions of both types of convection the following equation is valid:

$$Nu = \sqrt[3]{|Nu_{fc}^3 - Nu_{nc}^3|} \qquad (117)$$

For the application at solar collectors resp. solar modules, the following "over the thumb" formulas have been developed. Duffie *et al.* 1974 are suggesting a common convective heat transfer coefficient:

$$h_{nc+fc} = h_{conv} = 5.7 + 3.8w \tag{118}$$

h_{nc+fc} convective heat transfer coefficient in W m^{-2} K^{-1}

w wind speed in m s^{-1}

A separate treatment of the convective heat flows at the module front and at the module backside was carried out by Hart *et al.* 1982. They suggest the following heat transfer coefficients:

$$h_{F,conv} = 1.247 \sqrt[3]{(T_F - T_A)\cos\gamma_M} + 0.2658w \tag{119}$$

$$h_{B,conv} = 1.079 \sqrt[3]{(T_B - T_A)\cos\gamma_M} + 0.825w \tag{120}$$

$h_{F,conv}$ conv. heat transfer coefficient of the frontside in W m^{-2} K^{-1}

$h_{B,conv}$ conv. heat transfer coefficient of the backside in W m^{-2} K^{-1}

T_F surface temperature of the module frontside in K

T_B surface temperature of the module backside in K

w wind speed in m s^{-1}

γ_M elevation angle of the PV module

Flows oblique to the plates: Opposite to the cases with flows parallel or perpendicular to the plates, only a small amount of literature deals with flows oblique to the plate plated. Besides an exact knowledge of the surface structure of the plate, the accurate flow direction is also of importance. While in most cases meteorological data of that kind is not available, the equations (118)-(120) mentioned above are used.

7.2.4 Model Calculation

The aim of the model calculation is to determine the cell temperature T_C, which is computed first by calculating the heat flow input $\dot{Q}_{in}(E_{in}, \eta_{PV}(T_C))$ and them by carrying out a balance of heat flow input and heat flow outputs. While the heat flow outputs are a non-linear function of the surface temperatures of the front- and backside of the module (T_F and T_B) they have to be calculated in an iterative way.

For the stationary case $\dot{Q}_{in}(t) = \dot{Q}_{out}(t)$ (i.e., heat flow input = heat dissipation to the environment), heat flow output occurs by convection and by thermal radiation exchange with the sky (mainly module frontside) and with the ground (mainly module backside).

$$\dot{Q} = \dot{Q}_{nc} + \dot{Q}_{rad} \qquad (121)$$

$$\dot{Q}_{nc} = h_{nc_F} A_F (T_F - T_A) + h_{nc_B} A_B (T_B - T_A) \qquad (122)$$

$$\dot{Q}_{rad} = \sigma \varepsilon_F A_F (T_F^4 - T_{Sky}^4) + \sigma \varepsilon_B A_B (T_B^4 - T_G^4) \qquad (123)$$

$\dot{Q}_{nc}$ heat flow by natural convection in W

$\dot{Q}_{rad}$ heat flow by thermal radiation in W

h_{ncF} heat transfer coeff. for natural convection at the frontside in W m^{-2} K^{-1}

h_{ncB} heat transfer coeff. for natural convection at the backside in W m^{-2} K^{-1}

ε_F emissivity of the frontside of the module surface

ε_R emissivity of the backside of the module surface

$A_R = A_F$ surface area of the front- resp. back-side of the module in m^2

T_A ambient temperature in K

T_F temperature at the frontside of the module in K

T_B temperature at the backside of the module in K

T_{Sky} sky temperature in K

T_G ground temperature in K

While thick-film solar cells are becoming more equipped with features for total absorption of incoming irradiance (e.g., Morf 1990), the component being reflected to the outside from the backside of the cell does not have to be considered any more. For thin-film solar cells the irradiance component being transmitted through the module has to be subtracted from the incoming energy flow.

Commonly available simulation programs for the analysis of electrical circuits such as *SPICE* (Simulation Program with Integrated Circuit Emphasis) can be used to determine the value of T, $\dot{Q}$ and their parameter sensitivity when the thermal values are getting transformed converted into their analog electrical values, as shown in Table 7.6.

Table 7.6. Thermal – Electrical analogies

Thermal		Electrical	
Parameter	Unit	Parameter	Unit
temperature T	K	difference of electrical potential V	V
heat Q	J	electrical charge Q	Cb ≙ A s
heat flow $\dot{Q}$	W	electrical current I	A
thermal conductivity k	W m^{-1} K^{-1}	electrical conductivity σ	S ≙ A V^{-1}
thermal resistance R_k	W m^{-2} K^{-1}	electrical resistance R	Ω ≙ V A^{-1}
heat capacity $m \cdot c_p$	J K^{-1}	electrical capacity C	F ≙ A s V^{-1}

For clarity of the graphical illustration the thermal resistance is constituted by a series connection of plane resistors (see Fuentes 1985). This series connection is terminated by the heat transition of radiation, convection and conduction to the environment represented by nonlinear temperature dependent resistors grounded to the environments (sky temperature, resp. ambient temperature).

7.2.5 Validation of Thermal Modeling

During a post doc research visit of the author at the University of New South Wales in Sydney the upper models were validated experimentally. A BP 255 module was mounted in a thermally insulated way using an elevation angle of 30° towards North (see Fig. 7.23). Natural convection was undisturbed by obstacles. At the backside of the PV module (glass–EVA–Si–EVA–Tedlar®/Polyester/Tedlar® laminate) two Pt100 temperature sensors were attached. Measurement of ambient temperature was carried out with two independent thermometers in the shadow area of the module. Wind speed was measured by a precision air flow meter at the height of the module. Additionally short circuit current, open circuit voltage and the maximum electrical power output have been measured.

Differences between the simulated prediction and the actual measurement data were low for cell temperatures and exceeded a tolerance belt of ±1.5 K only when the actual wind speed was very different from the constant 2 m/s used in the simulation (see Fig. 7.24).

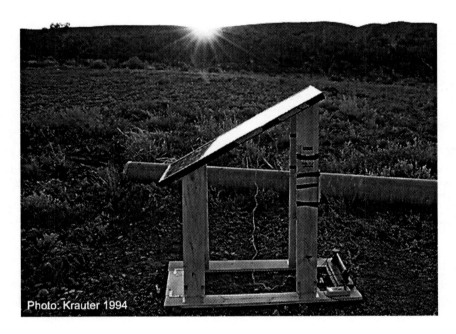

Fig. 7.24. Experimental set-up for the validation of the thermal and optimal modeling at a research station of the University of New South Wales at sunrise (Foulders Gap, Australia, 31° S, 140° E). Clearly visible are the high optical losses caused by the reflection of the flat incidence of the direct component of irradiance.

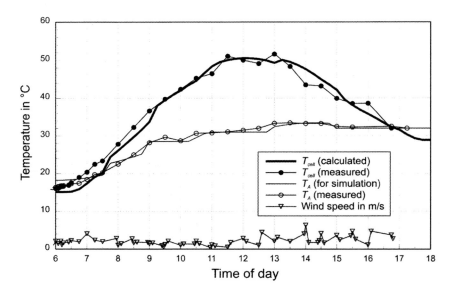

Fig. 7.25. Comparison of solar cell temperature simulation with actual measured cell temperatures during the course of a day (Foulders Gap, Australia, 21st of March 1994, clear sky).

7.3 Electrical Modeling

For the electrical modeling of the solar cells a so called "one-diode-model" was used (see, e.g., Green 1995 or Lund *et al.* 1987). This relative simple model serves to determine the PV output power as a function of irradiance, incidence angle, spectral efficiency, cell temperature and load.

7.3.1 Current

The short circuit current I_{sc} is determined by:

$$I_{sc} = \int_0^{\frac{\pi}{2}} \int_0^{\infty} E(\lambda, \theta_{in}) \cdot \tau(\tau_i(d_i, n(\lambda, T_i), \theta_i)) \cdot s(\lambda, T_C, \theta_i, E) \, d\lambda \, d\theta_{in} \quad (124)$$

I_{sc} short circuit current in A
$E(\lambda, \theta_{in})$ terrestrial solar spectrum as a function of incidence angle in W/m²
$\tau(\tau_i(d_i, n(T_i, \lambda), \theta_i))$ optical transmittance of the encapsulation
$\tau_i(d_i, n(T_i, \lambda), \theta_i)$ optical transmittance of the layer *i*

$n(T_i,\lambda)$	optical refractive index of the layer i
$s(\lambda,T,\theta_i,E)$	spectral sensivity of cell in A/W
E	irradiance in W/m²
d_i	thickness of the layer i in m
T_i	temperature of the layer i in K
T_C	cell temperature in K
λ	wavelength of irradiance in nm
θ_{in}	incidence angle on the module surface in °
θ_i	incidence angle on the layer i in °

The analytic modeling of the actual spectral sensitivity is quite extensive and requires a lot of material parameters which are usually not provided by the manufacturers. Therefore, the modeling uses measurement data for the spectral efficiency.

7.3.2 Other Electrical Parameters

The theory of photovoltaic conversion is well known and documented (e.g., Green 1995, 2000). While the actual modeling of the voltage is quite extensive and requires several internal parameters from the manufacturer that often are rather difficult to obtain by the PV system engineer, we shall use a "black box" approach, using the reference data given in the datasheet of a PV module at STC (V_{oc}, V_{mp}, I_{sc}, I_{mp}, TC (I), TC (V), TC (P)), and calculate the actual voltage by using the computed (or measured) temperature with the temperature coefficient. Voltage losses due to low irradiance are taken into consideration according to the logarithmic relation between irradiance and voltage. *TC (I)* is usually very small, but positive, due to the extended spectral photovoltaic conversion in the IR range. The form factor *FF* is also temperature dependent, therefore *TC (P)* is often higher than *TC(V)*.

7.4 PV grid injection

7.4.1 Modeling of Inverters

Inverters fed by PV modules are not operated at their nominal power all the time; depending on the actual weather conditions and time of day the input power may be significantly lower than nominal power. While the conversion efficiency of the inverters is decreasing when operated below nominal power, all possible points of operations have to be considered and a value has to be found which reflects the typical efficiency of an inverter for a long period of time. Such a value is the "European Efficiency" η_{Euro}, which pays account to typical European conditions and photovoltaic yields:

$$\eta_{Euro} = 0.03 \cdot \eta_{5\%} + 0.06 \cdot \eta_{10\%} + 0.13 \cdot \eta_{20\%} + 0.1 \cdot \eta_{30\%} + 0.48 \cdot \eta_{50\%} + 0.2 \cdot \eta_{100\%} \quad (125)$$

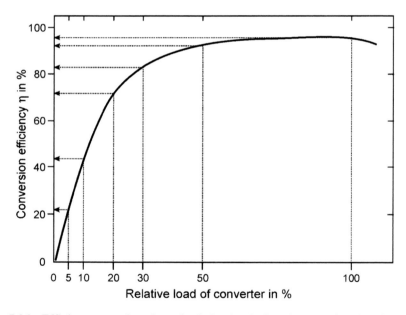

Fig. 7.26. Efficiency as a function of relative load of an inverter for electrical grid injection, together with the operation points relevant for the determination of the "European Efficiency".

For the example illustrated above, η_{Euro} would be 83.6 %. The efficiency characteristics of an inverter depends on the technology applied. To reduce low conversion efficiency at small loads, sometimes cascaded inverters are used: For small loads just one minor inverter is used, operating close to its nominal power, when the load increases, another inverter is switched on. Also the unit's self consumption losses, which are independent from the load, have to be considered. Grid injection occurs only when PV generation is higher than self consumption losses. To lower self consumption losses inverters are often automatically switched off during nighttime.

Another way to model the parameters of an inverter is presented by Schmid 1994. The efficiency characteristics of the inverter are modeled by three parameters:

$$\eta_{inv} = \frac{p_{out}}{p_{out} + k_0 + k_1 \cdot p_{out} + k_2 \cdot p_{out}} \quad \text{with:} \quad p_{out} = \frac{P_{out}}{P_{rated}} \quad (126)$$

k_0 loss coefficient for internal consumption
k_1 loss coefficient for voltage
k_2 loss coefficient for resistance
η_{inv} actual efficiency of the inverter
P_{out} actual power output of the inverter in W
P_{rated} maximum rated power output of inverter in W

An investigation of 35 different commercially available inverters by Jantsch et al. 1992 resulted in the average values of the following parameters: $k_0 = 0.02$; $k_1 = 0.025$; $k_2 = 0.08$.

7.4.2 Limiting Factors for the Design of PV Power Plants

In general, PV power plants can be operated at any irradiated location, even in space. An upper technical limit of the size of the modular construction does not exist, if the maximum insolation voltage of the PV modules is considered (in general: 600 V). In principle, two parameters limit the possibilities for reasonable extension:

- suitable locations
- electrical energy demand

Considerations for the integration of larger PV power plants into an existing electrical grid system and its limits for the extension of PV generating capacity can be found at Tzschoppe 1994. It is shown for the case of Germany that an extension of PV generating capacity by distributed PV power plants does not cause problems for the grid stability up to a GW scale. Difficulties will just occur when PV reaches 15–20% of total energy generation.

7.5 System Layouts

The energy supplied by renewable energy resources, such as the solar energy potential, is fluctuating due to climatic and weather conditions. Fig. 7.27 shows the energy- and frequency- distribution of solar irradiance for the example of Cologne, Germany. The contribution of the reference irradiance level of 1,000 W/m² for standard test conditions was very low, while most of the solar irradiance in Cologne occurs at levels between 180 and 500 W/m².

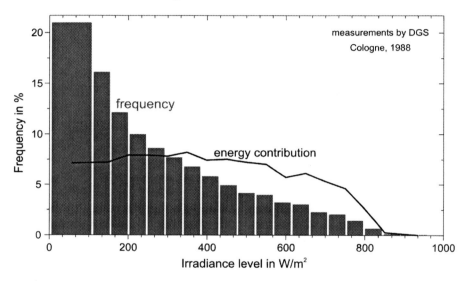

Fig. 7.27. Frequency and energy distribution of solar radiance for the example of Cologne, Germany in 1988 (Data by German Society for Solar Energy, Berlin, 1995).

While the irradiance levels of 1,000 W/m² occur rarely (at least in Central Europe), it is recommended to design the maximum power of the inverter so as not to be equal to the nominal power output of the PV generator under Standard Test Conditions (STC: E=1,000 W/m², AM 1.5, T_M = 25°C), but lowered by a reduction factor. The maximum electrical energy yield over a year for common inverters could be achieved for reduction factors between 0.65 and 0.8 for the case of Northern Europe; between 0.75 and 0.9 for the case of Central Europe and 0.85 to 1.0 for the case of Southern Europe and Northern Africa (see Macagnan 1992).

Extension of amplitude and dynamics – The efficiency versus load curves of most components in a PV system are characterized by a considerable reduction of conversion efficiency when operation conditions are away from standard conditions. In reality this occurs frequently (see Fig. 7.27), leading to significantly reduced system efficiency, amplified by the "series" connection of the components. This results in a higher amplitude of the fluctuations of electrical power output compared to the fluctuation and amplitude of energy input by irradiance. Therefore the system has to be able to handle an extended dynamic range over that provided by the dynamic range of the input (irradiance). Most components of the PV systems have a reduced conversion efficiency when operation conditions do not meet Standard Test Conditions.

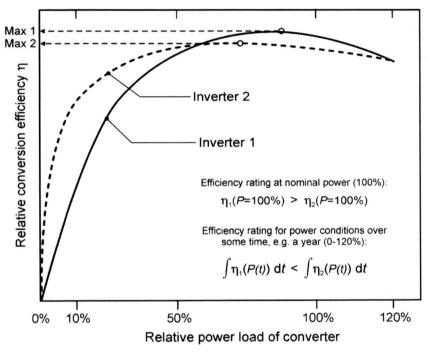

Fig. 7.28. Relative conversion efficiency only at a function of relative load for two different types of inverters: While inverter 1 offers peak conversion efficiency just at nominal load, inverter 2 offers high efficiencies over an extended range of loads.

Often system components can be optimized just for one parameter, for example, a maximum efficiency at rated power or a maximum average efficiency over a certain dynamic range. An enlargement of the dynamic range could hardly be made without a reduction of maximum efficiency, reducing also their average efficiency. Simply splitting one high power component into a number of smaller components ("cascading") can help. For low load levels just one component is in operation – for higher loads more comments are switched in parallel. Experiences of this type of operation were made for inverters (see above) and water pumps.

7.6 Electrical yield of a Reference System

Considering all loss-mechanisms described in the models above, the electrical yields given in Table 7.7 are achieved. The values given by Staiß 1997 are based on a PV reference system with multi-crystalline silicon solar cells and a module efficiency of 12%, installed in southern Germany (Munich).

Table 7.7. Electrical yield of PV systems (η_{STC} =12%) for grid injection in Germany

Power of the PV system	Power specific AC yield	Area specific AC yield
1–5 kW$_p$	784 kWh/(a kW$_p$)	94.1 kWh/(a m^2)
50–150 kW$_p$	859 kWh/(a kW$_p$)	103.1 kWh/(a m^2)
300–700 kW$_p$	887 kWh/(a kW$_p$)	106.4 kWh/(a m^2)

8
Energy Input by Dumping and Recycling

While often one type of literature offers an examination of the production process (Aulich et al. 1986, Hagedorn 1989), other literature carries out detailed investigation of electrical yields during operation (Krauter 1993c etc.). Further references offer product and yield improvements (e.g., Green 1995, Krauter et al. 1996a). An integral examination of the energy- and CO_2-relevance covering all impacts mentioned above in a total life cycle, including recycling, can be hardly found.

Literature for recycling is often based on other industrial sectors rather than PV but can be approximated by adopting equivalent material compositions.

8.1   Separation of Materials

Bruton et al. 1994 compared different methods to separate the material components of a PV module, regarding their recycling potentials:

- Mechanical delamination:
  leads to a destruction of the cells, only backside foils made out of Tedlar-Polyester-Tedlar could be torn off, by carefully heating it by a hot air pistol.

- Thermal delamination:
  Heating of the laminate to 200°C resulted in carbonization of the EVA, yet the silicon wafers rendered unusable. *German Solar* introduced a new process in 2003/04 based on thermal delamination, but is able to recover the solar cells - details of the process are kept secret.

- Chemical delamination:
  Different solvents have been tested, but they all resulted in a swelling of the EVA layer, which caused a breaking of the silicon wafer.
  Caustic soda solution was able to reduce the adhesion between EVA and the glass over the areas without cells, but insignificant reduction of adhesion was observed among cells, EVA and glass.

Finally nitric acid (HNO_3, corrosive, flammable) was able to dissolve EVA at a temperature of 80°C to 100°C and allowed a separation of the solar cells from the compound. The chemical reaction was as follows:

$$6 HNO_3 + CH_2 \Rightarrow 4 H_2O + CO_2 + 6 NO_2$$

The recycled solar cells have been cleansed by caustic soda und have been refurbished at BP Madrid. The recycled cells achieved the same open circuit voltage and the same fill factor as the reference cells, but the short circuit current was 20% less, which could be explained by a reduced surface structure. The cost of the recycling process was specified at 0.56 € per waver (respectively per cell) of a BP 255 module. An energy analysis has not been carried out, but according to the costs, the energy expense were assumed to be at least a factor of ten lower than for the production of a new cell.

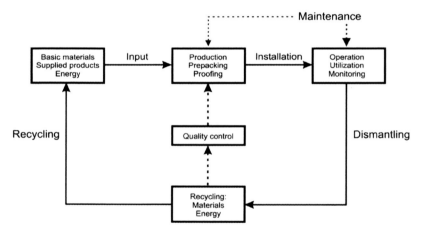

Fig. 8.1. Scheme for flows of energy and information within a life cycle of a PV system, considering recycling (energy flows are represented by full lines, flows of information by dashed lines).

8.2 Energy Input by Recycling

The energy saving potential by the use of recycled materials (secondary materials) for production is listed in Table 8.1. Table 8.2 by Hütte 2000 gives recycling quotes actually achieved, which are often quite considerable. Striking is the high saving potential by the use of secondary aluminum.

While the production of primary aluminum requires about ten times more energy than the production of primary steel, recycled aluminum has an energy requirement that is equivalent to new steel. This has a significant impact on the energy balance of PV modules equipped with aluminum frames. Unfortunately, data for the use of secondary aluminum is rarely available by the module manufacturers so that effect on energy balance could not be applied consequently in the calculations.

Examinations for PV modules made of amorphous silicon by van Engelenburg et al. 1995 have shown, that - after separation of the laminate - the module front glass is suitable as recycled white glass, if the module glass participation remains below 5%, or entirely as colored glass, independent from the participation of the module glass in the melt.

Table 8.1. Reduction of energy consumption for the production of materials by recycling

Material	Energy for primary material (new, from raw material)	Energy for secondary material (100% Recycling)	Reduction of energy requ. by recycling	Reference
Aluminum (sheet)	190 MJ/kg	30 MJ/kg	84.2%	Ebersberger 1995
Aluminum (min)	160 MJ/kg	12 MJ/kg	92.5%	Hütte 2004
Aluminum (max)	240 MJ/kg	20 MJ/kg	91.7%	Hütte 2004
Aluminum	142 MJ/kg	16 MJ/kg	88.7%	Frischknecht 1996
Steel (min)	16 MJ/kg	10 MJ/kg	37.5%	Hütte 2004
Steel (max)	27 MJ/kg	18 MJ/kg	33.3%	Hütte 2004
Cooper	95 MJ/kg	26 MJ/kg	72.6%	Wagner 1995a
Glass	10.8 MJ/kg	3.6 MJ/kg	66.6%	DB 1995
Paper	79.2 MJ/kg	18 MJ/kg	77.3%	DB 1995

Data basis for the calculations for cooper has been Wagner 1995a, Kaltschmitt 1995 and Fritsche 1989

Table 8.2. Recycling quotas of different materials

Material	World consum. 1983 in 10^6 t (Hornbogen 1994)	Recycl. quote Western World 1983 (Hornbogen 1994)	Recycl. quote Germany (Hütte 2000)	Recycl. quote Germany (Kaltschmitt 2003)	Recycl. quote Germany (Hütte 2004)	Recycl. quote Germany & EU (trade assoc., EU, aluinfo 2000)
Iron, Steel	650.80	29.2%	55%	40%	35.8%–49.7%	
Aluminum	16.35	26.7%	27%	63%	31%–35%	electr. eng.: 87% (D) 80% (EU) mech. eng.: 83% (D) 80% (EU) packing: 72% (D) 40% (EU)
Cooper		40.2%		56%	48.3%–54.6%	
Zinc	5.42	25.1%			34.9%–40.2%	
Lead	3.99	43.9%			45%–59.9%	
Sn	0.18	16.7%				
Glass			45%			
Paper			35%			90%
Plastics			10%			60%

9
Total Energy Balance

Taken into account are incoming and outgoing energy and material flows during production, operation, and dismantling (resp. recycling) of the photovoltaic power plant. The inherent energies of the material flows have to be considered in the energy and CO_2-balance.

To determine the inherent so called "grey energy" consumption, two methods are known: the energy input-output method based on macroeconomic input-output tables and the analysis of the subsequent processes, the "process-chain analysis." This work carries out the more accurate process-chain analysis as far as possible. In the case there is not sufficient data available for processes to be analyzed (e.g., services), an input-output analysis will be carried out.

9.1 Commutated Energy Expense

Definition:

"The commutated energy expense (CEE) pays account for the overall expense of energy, occurring at production, utilization, and liquidation of an object, respectively could be causative be related to it. This accordingly object-related energy expense is the sum of the commutated energy expenses for production (CEE_P), utilization (CEE_U) and liquidation (CEE_L)."

(Hagedorn 1990)

During the utilization of an object, energy consumption also occurs due to maintenance and supply of spare parts, if necessary. The environmentally sound disposal of a used-up object also may cause additional energy expenses. If components are reused or recycled, this could lead to a credit in the energy expense balance (negative consumption).

The CEE is an important value for the ecological classification of a product. To allow comparison with other products it has to be clearly visible how it will be utilized and for what has been paid account for.

It has also to be indicated whether the values are representing primary – or consumed energy expenses. If the CEE is given as primary energy expense, the actual conversion efficiencies of the power plants incorporated for the generation of the consumed energy have to be considered. Hagedorn suggests the following efficiencies which are also used in the calculations below:

Electrical energy: The average efficiency of thermal power plants in Germany is 0.35, modern power plants may reach 0.4.

Fuels: The consumption of energy carriers consumed as fuels are considered using an efficiency of 0.85.

NEC: Non-energy consumption (NEC) is the use of energy carriers as raw-materials (e.g., mineral oil for the manufacture of plastic) not as fuels, meaning not using their inherent energy content by burning. If no further data is given, NEC is taken into account by using an efficiency of 0.80.

9.2 Models for Energy Balances

When carrying out an energy balance, the calculation of the system's borders plays a very important role in the results of the analysis. The system borders, which are related to the methodology, are also called "vertical borders;" borders dealing with the definition of the problem are called "horizontal borders." Horizontal borders may limit a system to a specific part of a process chain. Vertical borders define which part of the economy supplies the vsystem with materials necessary, services and equipments. The following expenses and emissions have to be included in the energy- and CO_2-balances (Spreng 1995) [19]:

1. Direct energy expenses and resulting CO_2-emissions
during the process-step or by related activities referring to the system. Included are fuels, electricity, and internal consumption of the energy carrier which is being processed.

2. Indirect energy expenses,
including the energy of the materials consumed during the process: energy contained in the main equipment of the system, including the energy inside of all manufactured components as well as the energy spent directly during manufacture of these components; energy contained in the main equipments which are necessary for the production of materials and components,

---

[19] Spreng 1995 uses this method for energy balances, but the method can also be applied for CO_2 balances.

Total Energy Balance   179

including the energy consumption of the machinery used for production; the energy necessary for production of fuels and electricity. Sometimes the distinction between direct and indirect energy consumption is not easy. In such cases an energy flow diagram (Sankey-diagrams) gives clarity.

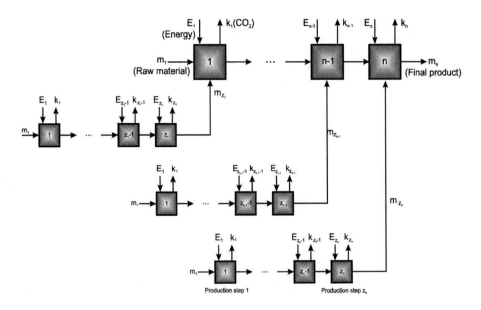

Fig. 9.1. General process-chain with one sub-chain.

9.3   Input-Output Analysis

This method is predominantly suitable for very large scale installations, e.g., for power plants. The relevance of the results is limited by the ability to resolve single sectors of the input-output table. For example, it is usually not possible to compare two concrete products in common practice. In parts energy- and CO_2-intensities for different products and services are already available in tabular form or will be made available soon due to the introduction of ISO 14000 *ff* (see Annex). For these components, in so far as they could be considered as a closed system, the expensive process analysis becomes obsolete (Spreng 1988).

The macroeconomic input-output analysis for the most important industrial sectors in Germany is given in Table 9.1.

Table 9.1. Relative material, energy and personnel cost shares for the ten most valuable industries (in terms of gross added value) in Germany (Reference: Statistical Yearbook of Germany 2003)

Sectors of industry	Cost participation in production in %				Gross added value in million €
	Material [1]	Energy	Personnel	Other [2]	
Street vehicle industry	67.1	0.6	19.9	14.3	61,685
Mechanical engineering	46.5	0.8	33.5	18.1	58,937
Chemical industry	47.8	2.9	21.8	27.8	41,432
Coking, crude oil processing	56.0	0.6	2.8	39.6	40,163
Food industry	59.4	1.7	15.1	21.8	31,401
Production of metal products	41.1	1.5	35.3	19.2	30,276
Prod. electricity generating and distrib. units	51.7	0.7	30.5	17.8	27,913
Production of plastic and rubber products	49.4	1.9	26.8	19.8	19,292
Production and processing of metals	58.4	6.3	23.0	16.7	16,585
Printing, publishing, duplication industry	25.3	1.0	41.1	29.7	15,477

[1] incl. provisioning products. [2] rent, interests, other services

The very raw definition of the material- and energy-flows of a product (e.g., electrical cable) is carried out according to the cost-participation of the incorporated industries (for the example of electrical cable: Electrical engineering, nonferrous metal production and chemical industry). More detailed values are given in the annex at Table A4.

9.4 Process Chain Analysis

The process chain analysis examines the different stages of the manufacturing process of the energy system in form of a micro-analysis. The system is described as a process chain, if final energy or an energy service is achieved from an energy resource passing through different process steps (Spreng 1995). For each single process of this chain, not just inputs and outputs have to be defined, but also all materials and equipments which are causing an indirect energy expense. Also by-products and other outputs leading to additional energy expenses have to be determined and have to be considered.

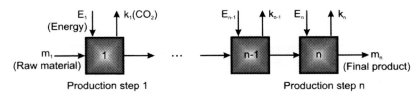

Fig. 9.2. Scheme of the energy-, CO_2- and material flows in a single process chain.

At first, the consumption of energy for operation has to be determined. This means, that all other energy inputs other than the input of the original energy to be transformed have to be settled (e.g., energy for internal transport). Then materials, equipment, operation substances and services have to be considered. To figure out their energy intensity, a second-order process chain analysis has to be performed. The energy balance has to be accounted for in the calculation of the original process. This procedure could lead to an infinite number of branches which have to be followed and analyzed; in practice only significant parameters are probed further than a second-order examination: A material "family tree" comes into existence (see Fig. 9.1), which branches up more and more. Each branch describes its own energy balance space, has its own history, which has to be examined separately in terms of cumulated energy expense and emissions. Even if an analysis could be limited to the first and second sub-chain, a process chain analysis requires a lot of work.

Aside from the selection of a typical manufacturing process, a lot of data is required, which generally is only obtained in close cooperation with the appropriate industries. It would be ideal if measurements could be carried out directly inside the production process. This requires, in most cases, an installation of measurement equipment at the specific location. For some materials and products, little or no data is available. Often the production processes are modified, for example, after the increase of energy costs in the years 1973 and 1978. Often, if the data is not based on recent examinations, old data shows significantly elevated energy-intensities which mean that the specific energy expense for the production of one ton of a specific material is assumed to be too high.

A production process includes different energy carriers: process heat (e.g., generated by coal of fuel oil) electricity or gasoline for the drive of engines. These energy chains have to be recalculated and adapted to a reference scale. Very often the primary energy expense is used. Thus for countries having a considerable participation of renewable energy for energy generation the equivalent primary energy consumption is quite theoretical and irrelevant. The way these energy chains are defined shows the inherent philosophy of its creators and their means of energy use, e.g., that of fossil fuel-based economies, which results that the production process is accounted for in coal ton units (or oil ton units), even if that energy carrier is never used in a process chain. – E.g., large producers of aluminum such as Norway or Iceland do not have a negligible use of fossil fuels for energy generation, so the conversion to coal units is just a loss in accuracy, while the applicable conversion factor is not constant and depends on the technology theoretically applied. While fossil fuel stock is limited and its application as a reference unit becomes conjectural over time and, on the other hand, the use of renewable energies increases in many parts of the world, it is reasonable to switch to a more sustainable reference unit.

Energy chain analysis is means for the creation of emission balances. The first step for that is the definition of the CO_2-emissions, which are related to the cumulated energy expense of a product. This is methodologically easy to do if a detailed inventory of the energy carriers is available. Also net emission balances can be created the same way as for the use of energy.

9.5 CO_2 Reducing Effects by the Use of PV

To examine this effect, a process-chain-analysis for a complete cycle of production, use, and recycling is carried out. In carrying out that, the results from the examinations above will be taken into account:

- Energy expense for production
- Transport and installation
- Yield of the PV generators during their lifetime
- Recycling

Due to a lack of industrial experience in recycling of PV modules, a recycling rate of 25% was taken for granted. This value represents a lower limit, other examinations such as Bruton et al. 1994 claim that values of up to 70% are feasible. Figure 9.3 shows the principal difference in terms of CO_2-emission resp. CO_2-reduction between conventional and PV power plants. While the PV power plant at the beginning of its life-cycle is showing some negative contribution to CO_2-reduction (but later it reduces CO_2-quite well). The contribution of conventional fuel-based power plants is always negative over the entire life cycle.

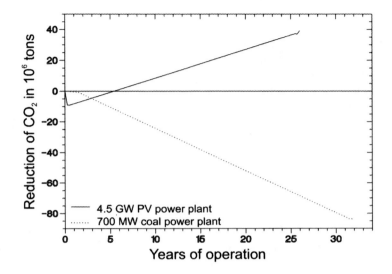

Fig. 9.3. Reduction of CO_2 during a life cycle of an 4.5 GW_p PV power plant operated in Germany in comparison to a 700 MW coal power plant with the same electrical yield (3.5 TWh/a).

9.5.1 Specific Emission Balance

Table 9.2 shows the commutated energy expense (divided in electricity, fuels, and non-energy consumption) and the CO_2-emission of a PV-power plant in Germany. The CO_2-balance in Table 9.2 was based on the sectors of energy expenses; for electricity the actual power plant composition was applied.

Table 9.2. Energy expense and CO_2-emission for the production of PV-power plants

Cell type	Electricity			Fuels		NEC [1]	
	Energy (kWh/kW$_p$)	CO_2 in D (kg/kW$_p$)	CO_2 in BR (kg/kW$_p$)	Energy (kWh/kW$_p$)	CO_2 (kg/kW$_p$)	Energy (kWh/kW$_p$)	CO_2 (kg/kW$_p$)
sc-Si	5,144	2,726	360	1,152	346	228	52.4
mc-Si	2,530	1,341	177	1,630	489	450	1,035

[1] Non-energetic-consumption: use of energy carriers as construction material. characteristics: electricity: 0.53 kg CO_2/kWh, fuels: 0.3 kg CO_2/kWh, NEC: 0.23 kg CO_2/kWh, input data from Hagedorn 1989, Wagner 1996, Höner 1997.

9.5.2 Effect of PV on Reduction of CO_2 Emissions in Germany

While the calculations given above indicate the possible reduction CO_2 via PV, the calculations below give an evaluation of the total possible reductive effect of PV applied in Germany.

The reduction effect of emissions is even higher if not just the actual composition of the electrical grid is taken into account, which emits on average 0.53 kg/kWh, but also the priority of policies and energy planning; e.g., energy consumption in Germany is stagnating at a high level, and additional power generating capacity will be used for substitution of existing power plants. The political will of all relevant parties in Germany is to reduce CO_2 emissions, in order to achieve that goal in an efficient way the power plants having the highest specific emissions are substituted first: the lignite and coal power plats, it should be also mentioned that these types of power plants are receiving considerable amounts of subsidies (see Table A7), which could be saved. If large amounts of PV or other renewable energies are applied, and the specific emission of the public electrical grid decreases in CO_2 output, lower emission during production of the PV systems could be achieved ("breeder-principle").

The first study for the potential for roof-mounted PV in Germany was carried out by Kaltschmitt 1993. The applied data basis was the German Census 1987 and was extrapolated to 1991 according to the statistic increase of buildings.

A good distinction has been obtained between pitched (slanted) roofs, flat roofs at residential buildings, commercial buildings, and also shed roofs used in factory halls due to good illumination. The latter one allows a maximum utilization of the roof area, while the mounting devices could be attached directly to the pediments.

Table 9.3. Reduction of CO_2 by PV power plants in Germany

Type of cell (lifetime)	Product. CO_2 (kg/kW$_p$)	Transp. CO_2 (kg/kW$_p$)	Install. dismant. CO_2 (kg/kW$_p$)	Recycling (25 %) CO_2 (kg/kW$_p$)	Yearly yield Energy (kWh/kW$_p$)	CO_2 (kg/kW$_p$)	Total balance CO_2 (kg/kW$_p$)
Single-cryst. (25 years)	3,124.0	52.9	15.0	- 781	770	- 408	- 7,791.5
Multi-cryst. (25 years)	1,933.0	58.2	16.5	- 483	770	- 408	- 8,677.3
Multi-cryst. (20 years)	1,933.0	58.2	16.5	- 483	770	- 408	- 6,637.3

Characteristic data:
electricity: 0.53 kg CO_2/kWh, fuels: 0.3 kg CO_2/kWh, NEV: 0.23 kg CO_2/kWh
PV inland transport:
350 km truck: 40.4 kg CO_2/kW$_p$, 50 km delivery van: 12.5 kg CO_2/kW$_p$
PV system weight:
single-Si: 330 kg/kW$_p$, multi-Si: 363 kg/kW$_p$, PV system efficiency (STC): 10%
Performance Ratio during operation: 70% (reflection, effect of temperature, spectrum, $\eta = f(G, \eta_{DC/AC})$
Irradiance: Ø 1,100 kWh/a, inp. data Hagedorn 1989, Wagner 1996, Höner 1997

Parameters limiting the installation of PV modules on roofs are considered as follows:

- The usable azimuthal sector is 90°, i.e., the permissible deviation from South is ±45°.
- Due to constructional constraints such as roof hatchways, ventilating shafts and chimneys, 20% of the roof area is not usable, at industry building an additional reduction of area 15% is assessed.
- Other restriction such as shadowing (10%), protection of historical buildings and monument (5% just are slanted roofs) is further lessening the potential.
- For flat roofs the maximal module cover is 33%

In total 16% of the total roof area of slanted roofs and 25% of flat roofs is adequate for the application of solar energy systems. This corresponds to module area of 800 km^2 for Germany.

The share of residential and nonresidential buildings is about the same. In this area 75.3 TWh of PV electricity could be generated, which stands for a reduction of 40 million tons of CO_2 under actual conditions and corresponds to 5.6% of the total CO_2 emissions. Considering the emissions for PV-production under actual conditions, this value is reduced by 20%. If the usable open sites are also taken into account, a module area of at least 2000 km^2 could be derived, even if only the excessive agricultural area is considered. (Staiß 1996). This results in a yearly electricity yield of 211 TWh taking into account the higher efficiency of larger power plants on open sites. That corresponds to 46% of the net electricity consumption of 460 TWh (1993). The electrical supply network and the load profiles have to be adapted in terms of additional storage and energy exchange. By considering just the existing storing capacities only, the possible participation of PV is reduced to 8.7%, corresponding to a module area of 400 km^2. Considering net stability requires a permanent participation of 25% of conventional power plants, participation of PV power plants is just 6.5%. If adequate storage is available, allowing to store 2% of the entire energy, PV contribution can be 20% (see Beyer et al. 1990, Staiß 1995), corresponding to a reduction of CO_2 by 47.7 Mio tons, i.e. 6.7 % of the German CO_2 emissions. The assumptions made for the calculations are conservative and restrictive, e.g., just sites having a maximal loss of 10%, compared to optimal module orientation, have been taken into consideration. Utilization of building facades and open sites at residential estates and at traffic structures have not been considered either.

9.5.3 Variation of Location

While the specific electrical energy requirements do not vary notably for most of modern manufacturing facilities of PV components all over the world, the specific CO_2 emissions depend very much on the power plants (nuclear, hydro, fossil etc.) producing the electricity to operate production facilities of PV and system components. The CO_2 intensity of electrical power plants and of national electrical grids may vary considerably (between 17 and 1,140 g of CO_2/kWh_{el}), as can be seen in Table 9.4.

Remarks:
Due to a considerable increase of wind and solar energy contribution in some countries (e.g., Germany, Spain) during the last years, the participation of "other renewable energies" in Table 9.4 has changed (e.g., in Germany from 1.9% to 6.7% in 2005; Spain increased its installed wind power capacity from 75 MW in 1994 to 8,263 MW in 2004). More specific data for Germany is given in Table A7 in the Annex.

Total Energy Balance   187

Table 9.4. Composition of power plants for electricity generation in different countries (data by Schaefer 1993, Mauch 1995, Tahara et al. 1997, VDEW 1998/99 - published in 2002*), see also Table A7

Country	Fossil fuels	Nuclear power	Hydro power	Other Renewable Energies	CO_2 intensity of electricity (g/kWh)
Greece *	89.3 %	0 %	10.5 %	0 %	1111
Germany*	62.4 %	31.2 %	4.5 %	1.9 %	517
Netherlands*	94.5 %	4.3 %	0.1 %	1.1 %	442
Japan	61.1 %	28.2 %	10.5 %		439
Great Britain*	71.3 %	25.1 %	2.3 %	1.3 %	438
UdSSR (former)	74.7 %	12.4 %	12.9 %		
Spain*	55.7 %	28.3 %	13.8 %	2.2 %	393
Venezuela	43.0 %	0 %	57.0 %		
Brazil	6.0 %	0.8 %	93.2 %		70
France*	9.7 %	75.0 %	15.3 %	0 %	61
Sweden*	4.4 %	46.6 %	47.0 %	2.0 %	34
Norway*	0.7 %	0 %	99.3 %		16
Ghana	0.2 %	0 %	99.8 %		
Iceland	0.1 %	0 %	99.9 %		15
Mix of aluminum exporting countries					139
Mix of copper exporting countries					572

For example, if a PV system is produced in Brazil instead of Germany, then the specific CO_2 emission is reduced from 0.517 kg/kWh to 0.07 kg/kWh for electricity consumption. Therefore the CO_2 emission for production of a PV system is also lower, on the other hand the CO_2 reduction effect during the operation phase is also lower (considering PV grid-connection and actual

188   Solar Electric Power Generation

state of the local grid, see Table 9.5)[20]. Ideally, PV systems are produced where CO_2 emissions for production are lowest, and are installed and operated where they are able to avoid maximum CO_2 emissions (see Table 9.6). The effect of transportation, if it is carried out by sea cargo, is quite low.

Table 9.5. Example of a CO_2-balance for PV systems in Brazil connected to the grid

Type of cell (lifetime)	Produc. CO_2 (kg/kW$_p$)	Transp. CO_2 (kg/kW$_p$)	Install, removal CO_2 (kg/kW$_p$)	Recycl. (25%) CO_2 (kg/kW$_p$)	Yearly yield Energy (kWh/kW$_p$)	CO_2 (kg/kW$_p$)	Total balance CO_2 (kg/kW$_p$)
Single-cryst. (25 y)	758.0	158.7	30	-189.50	1,225	-85.8	-1,386.6
Multi-cryst. (25 y)	769.5	174.6	33	-192.40	1,225	-85.8	-1,359.1
Multi-cryst. (20 y)	769.5	174.6	33	-192.40	1,225	-85.8	-930.1

characteristics:
electricity: 0.07 kg CO_2/kWh, fuels: 0.3 kg CO_2/kWh, NEC: 0.23 kg CO_2/kWh
PV inland transportation:
1,050 km Truck: 121.2 kg CO_2/kW$_p$, 150 km delivery van: 37.5 kg CO_2/kW$_p$
PV system weight: single-Si: 330 kg/kW$_p$, multi-Si: 363 kg/kW$_p$
PV system efficiency (at STC): 10%
Performance-ratio at operation: 68% (optical reflection, temperature effect, spectrum, $\eta=f(G)$, $\eta_{DC/AC}$)
Irradiance: Ø 1,800 kWh/a, inp. data Hagedorn 1989, Wagner 1996, Höner 1997

The fact that many components, semi-finished products and raw materials are imported makes the examination more difficult; beside the composition of the plant parks for electricity generation at the site of production (and its specific emission), the situation in the countries from which the goods are imported has to be taken into account. E.g., in Germany 50% of the primary

---

[20] Due to the increase of electricity consumption, Brazil will require further power plants: While there are not any adequate locations for locations for large hydro power plants left, alternative scenarios will be developed, changing the energy mix and the CO_2 intensity of Brazil.

(new) aluminum is imported, so the composition of the power plant park at the countries where the material is imported has to be taken into account.

While most aluminum exporting countries have a high level of hydro power participation, the equivalent "primary energy consumption" for aluminum actually used and processed is just 169 MJ/kg. In literature, often only the nation's power plant composition is considered, leading to an overestimated "primary energy consumption" of 227 MJ/kg. An overview of the composition of power plant park and CO_2-intensities of the main aluminum and cooper ore exporting countries is given in Table A18 in the Annex. For operation in Germany (yearly irradiance in Berlin on an optimal tilted plane is 1,050 kWh m^{-2} a^{-1}, the electrical energy yield for a 1 kW$_p$ PV system is 770 kWh$_{el}$ a^{-1}, in Rio de Janeiro (yearly irradiance 1,750 kWh a^{-1} m^{-2}) average electrical power output of the same 1 kW$_p$ PV system is about 1,138 kWh$_{el}$ a^{-1}. A corresponding issue as for the different locations of production occurs on the sites of application: Substituting a small Diesel-generator (0.9 to 1.05 kg CO_2/kWh$_{el}$) with a PV system could avoid 0.85 to 1 kg CO_2/kWh$_{el}$ of emissions, while an electrical grid connected PV system in a "clean" grid (e.g., Brazil at 0.07 kg CO_2/kWh$_{el}$) will not considerably reduce CO_2, especially if the PV system was produced using electricity from a "dirty" grid. In such a case the effect in terms of CO_2-reduction might even be negative.

Table 9.6. CO_2 reduction by PV systems as a function of location production and operation

Production site: Installation site:	Germany (Berlin)	Brazil (Rio de Janeiro)
Germany, grid-connected (Berlin, 52°30' N; 13°22' E)	7,792 kg/kW$_p$ (sc-Si, 25 a) 8,677 kg/kW$_p$ (mc-Si, 25 a) 6,637 kg/kW$_p$ (mc-Si, 20 a)	10,124 kg/kW$_p$ (sc-Si, 25 a) 9,805 kg/kW$_p$ (mc-Si, 25 a) 7,765 kg/kW$_p$ (mc-Si, 20 a)
Brazil, grid-connected (Rio de Janeiro, 22°53' S; 43°12' W)	-1,009 kg/kW$_p$ (sc-Si, 25 a) 162 kg/kW$_p$ (mc-Si, 25 a) -267 kg/kW$_p$ (mc-Si, 20 a)	1,387 kg/kW$_p$ (sc-Si, 25 a) 1,359 kg/kW$_p$ (mc-Si, 25 a) 930 kg/kW$_p$ (mc-Si, 20 a)
Brazil, off-grid (replacement of diesel generator by PV plant)	24,408 kg/kW$_p$ (sc-Si, 25 a) 25,372 kg/kW$_p$ (mc-Si, 25 a) 19,860 kg/kW$_p$ (mc-Si, 20 a)	26,805 kg/kW$_p$ (sc-Si, 25 a) 26,570 kg/kW$_p$ (mc-Si, 25 a) 21,058 kg/kW$_p$ (mc-Si, 20 a)

Reference data for Table 34:
Electricity in Germany (grid): 0.53 kg CO_2/kWh,
Electricity in Brazil (grid): 0.07 kg CO_2/kWh,
Electricity in Brazil (off-grid, via diesel generator): 0.9 kg CO_2/kWh,
Fuels: 0.3 kg CO_2/kWh, Non-energy consumption: 0.23 kg CO_2/kWh
Weight of a PV system: sc-Si: 330 kg/kW_p, mc-Si: 363 kg/kW_p,
PV system efficiency (STC): 10%
PV inland transport in Germany: 350 km, truck: 40.4 kg, 50 km van: 12.5 kg CO_2/kW_p (sc-Si), 58.2 kg CO_2/kW_p (mc-Si);
PV national transport in Brazil for sc-Si: 1050 km truck: 121.2 kg CO_2/kW_p, 150 km delivery van: 37.5 kg CO_2/kW_p, for mc-Si: 174.6 kg CO_2/kW_p (off-grid: 2 x transport and installation)
Transport Germany-Brazil: 10,000 km via cargo-ship: sc-Si: 31 kg CO_2/kW_p, mc-Si: 34.1 kg CO_2/kW_p
Performance-Ratio during operation: 70% in Germany, 68% in Brazil
(losses by reflection, temperature, spectrum, $\eta=f(G)$, $\eta_{DC/AC}$)
Irradiance: Ø 1,100 kWh/a in Germany; Ø 1,800 kWh/a in Brazil.
Recycling quote: 25%.
Data for other energy expenses by Hagedorn 1989, Wagner 1996, Höner 1997.

Due to different load requirements, the composition of generating sets and its specific CO_2 emissions in an electrical grid may vary during a day. Peak loads (e.g., in Brazil during weekdays between 5 p.m. and 10 p.m. the load factors are reaching 40%) are often served by fossil fuel driven power plants, which are increasing the average CO_2 emissions (and also their value for substitution) during these times. Unfortunately, PV power output does not match these peaks in the Brazilian interconnected grid system as a whole. On a local level, however, some grids are showing a good match between power demand and PV generation (Florianópolis – CELESC). PV can have a greater value for the utility in terms of grid-support in such instances. In recent years limitations of electrical power supply in Brazil have often not been given by the maximum rated power output of the hydro generators, but only by the amount of water stored in the dam, so the hydro generators are running at reduced power and can adapt to peak demand. PV can also contribute here by displacing or offsetting water levels in dams for the use during peak demand. Furthermore, PV and hydro generation can be regarded as complementary during a seasonal basis since dam water levels reach critical low values which coincide with higher solar irradiation levels in summer in the Northeast of Brazil. In Germany, daily peak loads occur earlier and could be matched in part by photovoltaics. The additional power plants operating at peak loads are hydro storage, natural gas, oil and mixed fuel powered plants. Due to the dynamic trading of electricity between companies and countries, especially during peak hours, and the difficulty of figuring the kind of energy used to fill the storage dams, an accurate calculation of CO_2 balance for peak conditions is quite extensive and will not be presented here.

10
Optimization

The objective is the maximization of the yearly electrical energy yield. This task is carried out by examining the effect of the actual environment in terms of irradiance, optical reflections, ambient temperature, wind as well as the performance of the different components and their interactions.

Optimization of cell-reaching irradiance:
Taking into account direct and diffuse irradiance (considering also angles of incidence spectrum and polarization) and the albedo.

Solar-electrical converter / solar cell:
Technology, feasibility, availability, cost-effectiveness, and environmental soundness. Effects of the operation environment (spectrum, incidence angles, temperature, shadowing, albedo, contacts, wiring) on the power output.

Solar modules, PV generator:
Implementation of the PV generator in an (e.g., urban) environment considering new concepts for installation and for an increment of the yield by control of optical, thermal and electrical parameters including the interface PV-module - environment (e.g., heat transfer, optical transition). Also concepts for the architectural integration at minimal costs by substitution of facade- or roofing elements (e.g., solar roof tile with an integrated concentrator, see Wenham et al. 1997; co-generation, see Bazilian 2002)

DC-AC conversion, grid feeding devices:
Comparison of different concepts (e.g., cascaded inverters, string-orientated inverter, module integrated inverter) taking into account the cost-benefit ratio.

10.1 Improvement of Irradiance on a Solar Cell

10.1.1 Improvement of Irradiance by Tracking the Sun

Irradiance on a solar generator can be improved by tracking the sun's path from sunrise to sunset. In the case of Berlin, Germany the yield increases by 25–30% in summer and by 0 –10% in winter. Most single-axis tracking devices follow the sun's path from east to west (azimuthal tracking) at a fixed inclination. Two axis tracking devices also follow the seasonal change in the elevation of the sun's path. An example of a two-axis tracker in comparison to a fixed PV-generator is shown in Fig. 10.1. The costs for a tracking device are considerable. In many cases the additional yield is not able to compensate for the extra costs (Ertürk 1997). In some parts of the Third World (e.g., India) a high cost-efficiency was achieved by manual tracking (e.g., every 3 hrs), due to low costs of labor.

Fig. 10.1. Fixed mounted PV modules and 2-axis tracking of a 6.6 kW$_p$, PV generator (Hans Grohe, Offenburg).

10.1.2 Improvement of Cell Irradiance by Reduction of Optical Reflection

At a fixed (non-tracking) PV installation the incidence angle of solar irradiance is rarely perpendicular so refection losses increase in comparison to the possible minimum (at plane surfaces). At unfavorable conditions, for example a facade-integrated PV system in low latitudes, the reflection losses of direct irradiance may amount up to 42% (Krauter 1993c).

Reflection losses can be reduced by changing the surface properties (structuring), a better matching of the refractive indices of the incorporated optical layers above the solar cell or by additional anti-reflective coatings.

10.1.2.1 Structuring of the PV Module Surface

A suitable structure for an optical surface, which prevents irradiance reflected from a structure flank of the surface to be lost, allows it to re-enter (in major parts) the PV-module when hitting the neighboring affronted flank of the structure.

In order to obtain indications about the entire module performance a three-layer flat-plate simulation program was modified to also consider V-structured surfaces. The results for three different grooving angles are presented in Figure 10.2 that shows the relative irradiance on the cell as a function of the incidence angle for different grooving angles at 90°; 120° and 150°. Over a relatively large range of angles of incidence an increased level of irradiance on the solar cell (compared to the planar surface) can be noted.

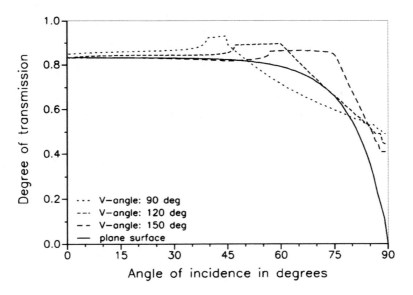

Fig. 10.2. Optical transmission of an overall V-shape structured PV module surface (no selective structuring applied) for different V-grooving angles in comparison to an un-structured (plane) surface.

10.1.2.2 Selective Structuring

A significant advantage obtained by selective, partial structuring of the cover is its deflection capability. For example, for a 1.5 mm wide front contact ("bus bar"), the irradiance onto a plane surface over it is useless, but the V-structure can direct it (completely within an incidence angle area of ± 5° to the normal of the plane) onto active areas. As structuring of the front layer is only done partially, the value for transmittance of the whole area is formed by the weighted average of structured and unstructured areas. The gain to be expected is about 60% of the front contact and busbar area (depending on the profile and reflectivity of the contacts) and 80% to 95% of the space area between the cells.

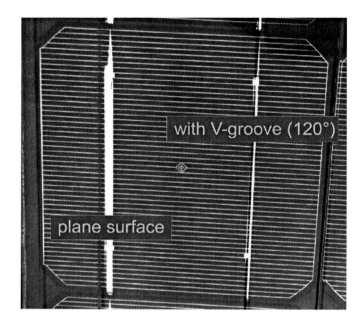

Fig. 10.3. Left: Conventional, plane module surface. Right: Optical reduction of the width of a contact stripe (bus bar) by application of a V-groove on the surface above it.

By applying a partial structure as shown in Figure 10.4 above photovoltaically inactively areas only (for example above contact grids (rays A) and cell spacing (ray B)) incoming radiation can be directed onto active ones.

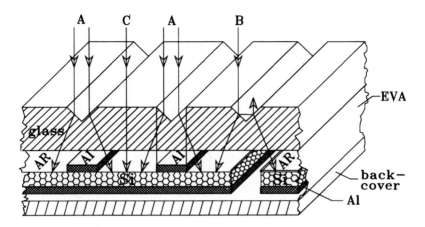

Fig. 10.4. Partial structuring of module surface by V-grooves over photovoltaic inactive areas.

At the air-glass interface the optical transmission increases from 95.7% at a plane surface (see ray C), to 99.7% at the V-shape-grooves (see rays A, B), due to the fact that reflections are not being completely lost, but reflected from the incidence surface to the opposite surface of the V-groove (unpolarized radiation, normal incidence, V-grooving angle 90°; n_{glass} = 1.52 at λ = 550 nm). Since the incidence angles onto the lower layers becomes larger when radiation was refracted by a groove, the gain for the cell-reaching radiation is reduced.

At an incidence angle of 80° the transmission at the incidence surface increases from 60.9% (plane surface) to 95.4%, but the opposite surface of the groove remains passive (as no radiation is reflected onto it) therefore the weighted gain (referenced to the projection area) is reduced. In case of soil depletion in the structure, no shadowing of PV active areas will occur. Fig. 10.3. shows clearly the effect of the optical reduction of inactive cell areas, such as the bus bars, by structuring. For incoming irradiance just about 30% of the active contact area is "visible", thus reducing the bus bar shadowing loss by 70%.

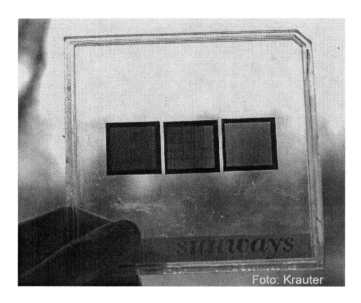

Fig. 10.5. Bi-facial V-structured silicon wafers.

Aside from improvements via structuring the module surfaces, the absorption properties inside the solar cell can also be enhanced by structuring. Beside the known pyramid structure (see Green 1995), V-groove structures on both sides (grooving direction orthogonal to each other) are also applied (see Figure 10.5). Besides less material used for the solar cell, also efficiency should be increased.

10.1.2.3 Improved Matching of the Refractive Indices of the Module Encapsulation Layers

Using the optical model presented above as a simulation for the optical system consisting of front glass, EVA, anti-reflective coating, and silicon solar cell, a parameter variation leads to the following results: A better optical matching of the two upper layers (glass and EVA) allows an increase optical transmittance (and consequently electrical yield) by 3.2% for materials with ideal properties, and 1.9% for real materials (see Krauter 1993c). Figure 10.6 shows the transmittance for perpendicular, unpolarized irradiance for a variation of the parameters n_1 (refractive index of front layer) and n_2 (refractive index of second layer). Also the transmittance of a real PV module (PQ 40/50) is shown.

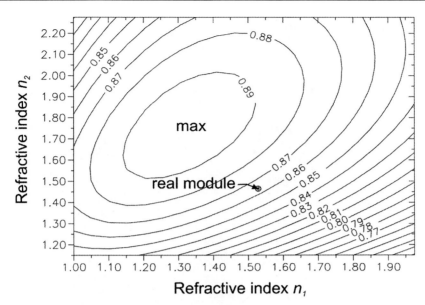

Fig. 10.6. Optical transmittance of a module encapsulation as a function of the refractive index of the two upper cover sheets (optical slabs) for perpendicular incidence of irradiance ($\theta = 0°$).

10.1.2.4 Additional Anti-Reflective Coating

To achieve an optimal matching of the module surface with the air, an additional anti-reflective coating (ARC) should have a refractive index of $n=1.3$. Unfortunately, no solid material has got such properties. Nevertheless, experiments using a liquid film as an AR-layer have been carried out: Using water ($n=1.33$) as an AR coating resulted in a 2% increase of the short-circuit current (Krauter 2004).

However, optically thin layers (e.g., $\lambda/4$), that allow a reduction of reflection losses even for relatively refractive indices, have been successfully applied at solar thermal collectors and are commonly used at high-quality optical equipment (camera lenses, binoculars, etc.). First application tests for PV modules (incl. evaluation of durability and cost-benefit ratio) are on the way.

10.2 Reduction of Expenses for Mounting

A new mounting system presented below (Fig. 10.7) allows to install PV modules without tools just using simple stainless steel cramps (see Fig. 10.8). The bending stress of the support structure fixes the components together and secures a tight mount of the solar module.

Fig. 10.7. Reduction of mounting cost by the Solbac®. support structure.

The support structure is made of Eternit® (see Fig. 10.7) and is very suitable for flat roofs, while the fixation to the roof could be achieved just by a pebble stone filling without the necessity of roof perforation. Substitution of the conventional metal support structure by Eternit® allows to reduce primary energy consumption in the vicinity of 90%. However, the material contains asbestos, so dicing should be carried out with great care, while the dust from sawing is carcinogenic. Other developments are using recycled plastic materials (e.g., PP, PE).

Fig. 10.8. Mounting cramp made of stainless steel (V2A) of Solbac® for framed PV modules.

10.3 Substitution of Building Components

Substituting a conventional component of a building –, e.g., roof tile, window or facade – with an adequate solar energy component, results in significant material, energetic and financial cost savings, while the structural components, such as framing and glass sheets, do not have to be considered as part of the solar generator in the balance.

10.3.1 Solar Roof Tiles

Figure 10.9 shows a roof covered by solar roof tiles made out of colored acrylic glass. In each of the tiles 24 single-crystalline silicon solar cells are integrated, the wiring among the solar roof tiles is carried out by a plug system. A similar system, but consisting of amorphous silicon solar cells, is manufactured by the Swiss company Atlantis. The costs are around 4.30 €/m² and thus a new roof a system of such type is considerably less expensive than a conventional roof with an additional mounted PV generator.

Fig. 10.9. A roof covered by solar roof tiles manufactured by the Swiss company Newtec (Wildnau).

Fig. 10.10. Single solar roof tile from Newtec, Switzerland.

10.3.2 Solar Facades

While solar facades are less favorable from the technical point of view (e.g., 30% less irradiance in Germany), they are very attractive in terms of visual appearance. Companies and institutions which want to express their environmental competence and conscience, (e.g., energy suppliers, construction industry, banks, insurances) often apply PV facades to call public attention and recognition. Beside the losses by lower irradiance, higher reflection losses occur also. In locations close to the equator, where the sun's elevation is high, reflection losses may amount 42% of the incoming irradiance (see Krauter 1994a). Nevertheless, the installation of a PV-facade may end up less expensive than a roof mounted system plus an expensive high-grade conventional facade: Facades made from granite, marble or special glass end up at 750 to 2,000 €/m² and have at least the same price as PV. In this case, PV facade elements can be applied without extra costs in comparison to conventional facades. This means that PV electricity from these facades is almost for free. The application of PV facades also triggers the imagination and desires of architects: Colored solar cells, granite-like crystalline structures, triangular and hexagonal framing, even mosaic-type of PV-module arrangements came into appearance over the recent years (see Figure 10.11). Different color appearance is achieved by the variation of the anti-reflective coating the solar cell (see Mason et al. 1995).

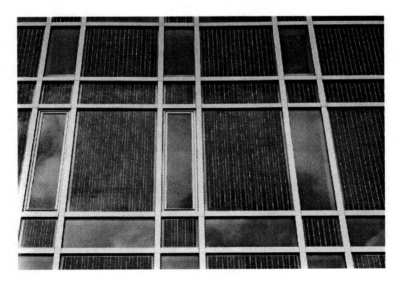

Fig. 10.11. PV-façade equipped with multi-crystalline PV modules and windows in a in chess-board type of arrangement (seen at "Stadtwerke Aachen").

Fig. 10.12. PV-façade at the Ökotec-building in Berlin-Kreuzberg (right side) in conjunction with a historic façade from the "Gründerzeit" (left side).

Fig. 10.13. PV-facade with PV modules as window shading elements to reduce air-condition load and glare (seen at Solarworld in Freiberg, Saxony).

10.4 Thermal Enhancement of PV Modules

10.4.1 Real Operating Cell Temperatures Under Tropical Conditions

To know more about real operating behavior under tropical climatic conditions, specifically about cell temperatures and the output power (at MPP), a module (M55 from SSI) was tested at the PV Labs of the UFRJ in Rio de Janeiro (22°54' S; 43°13' W), Brazil during an equinox (9/22/1994). The components of irradiance (horizontal global, direct and diffuse) during that day have been recorded, as shown in Figure 10.14.

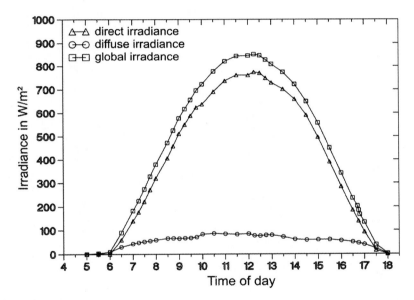

Fig. 10.14. Horizontal global, direct, and diffuse irradiance at Equinox (9/22/94) in Rio de Janeiro, Brazil (22°54'S; 43°13'W).

Figure 10.15 gives the measured temperature values at an M55 module during that day at almost no-wind conditions. Elevation angle of the module plane was 23°, facing North. The maximum power point was tracked manually by a variable resistor using a digital ammeter, a digital voltmeter and a calculator. Thermal and electrical measurements could be carried out just after 6:45 a.m., due to access-restrictions of the site, and not later than 4:45 p.m., due to partly shadowing of the generator by a neighboring building. The measured electrical energy output of the module was 300 Wh, while a constant cell temperature of 25°C, as requested in the Standard-Test-Conditions (STC), would have caused a 12% higher yield (see Figure 10.16).

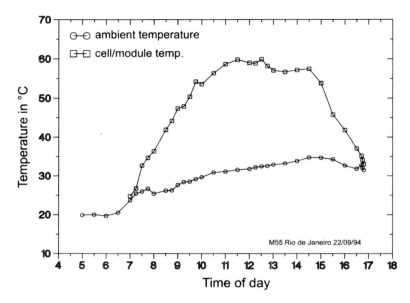

Fig. 10.15. Measured ambient-, and module temperatures at a M55-module during 9/22/1994 in Rio de Janeiro, Brazil, at no-wind conditions.

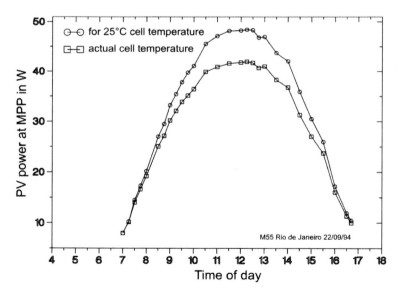

Fig. 10.16. Measured electrical power output under real conditions in comparison to the theoretical power output at a constant cell temperature of 25°C ("Maximum-Power-Point" conditions for both curves).

10.4.2 Preliminary Work for the Reduction of Temperatures in PV Modules

Research on an increase of PV-efficiency and electrical power output by means of a reduction in operating cell temperature has been carried out by the author since 1989 in a Ph.D. thesis (see Krauter 1993c).

The energy consumption of an active cooling system would not be compensated by the gain in increased energy generation, at least for small systems. Operational temperatures were kept at low levels by mounting the module on a water-filled tank. This allowed for an effective reduction in operating cell temperatures without spending any energy for refrigeration. The water virtually soaks up the heat flow generated by the module. Due to the high thermal capacity of the incorporated water ($C_{p,water}$= 1,254 kJ K^{-1}) the temperature increases gradually (see also results below). Also, the peak temperature is shifted from midday to afternoon. The principle was proven and validated with different prototypes in Europe and in Africa built over previous years.

History – The first cooling device which followed the "cooling by an extended heat capacity" concept was built in 1992. The tank was integrated into the original framing of an M55 PV module by SSI (former Siemens Solar Industries, now Shell Solar) with a volume of 12 liters, so it could be used with conventional mounting. This prototype provided a 2.6% increase in the daily electrical energy yield. Subsequent tests, which utilized latent heat storage material (sodium sulphate), showed significantly better results but caused severe corrosion (see Krauter 1993c).

10.4.3 Development of a Thermally Improved Prototype

The second prototype built in 1994 had a much larger water tank which served also as the module's foundation, its stand and mounting structure (TEPVIS – Thermal Enhanced PV module with Integrated Standing). It was tested with an M55 in Berlin (see Figures 10.17, 10.18, and 10.19) and showed an energy gain of up to 12%, and with PQ 10/40 devices in Bulawayo, Zimbabwe and proved an increase of 9.5%. The gain in Zimbabwe was lower due to reduced water circulation and more stratification (the upper part of the tank got considerably warmer than the lower one). The inclination of the module plane in Zimbabwe was much lower (20 degrees, according to the latitude of Bulawayo), thus reducing the thermo-siphon effect. The device in Berlin was also equipped with an additional plate inside the tank, in parallel to the module at a distance of 12 cm, forming a kind of chimney and thus enhancing circulation ("Onneken's separator"). Nevertheless, the

PV conversion efficiency of the cooled module was noteworthy above the reference module during at least 95% of the day.

To eliminate possible measurement errors which may have been caused by the differing electrical properties of the systems, all modules (reference and cooled ones) were interchanged and retested. The modules had been operated continuously at the Maximum Power Point (MPP) by manual tracking of an ohmic load together with power metering. The application for the new I-SHS was carried during the first quarter of 2002.

Fig. 10.17. TEPVIS (Thermal Enhanced PV with Integrated Standing): "Proof-of-principle" prototype made of acrylic showed a gain in electricity yield of 11.6% (Berlin, Germany, 1995).

10.4.4 Test of Prototype

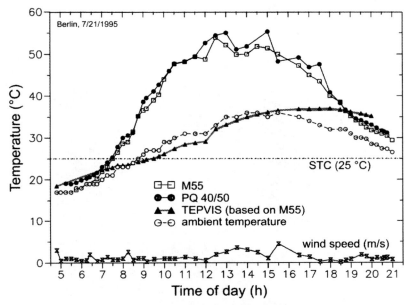

Fig. 10.18. Temperatures of TEPVIS in comparison to two conventional PV modules (PQ 10/40, M 55), all operated at Maximum Power Point (MPP), with ambient temperature and wind speeds for a clear day in Berlin, Germany (7/21/1995).

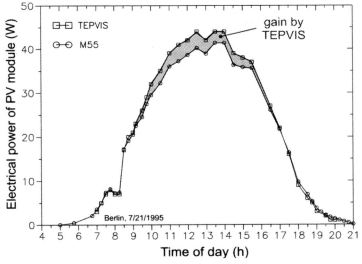

Fig. 10.19. PV power output of TEPVIS in comparison to a conventional M55 with conventional mounting, both operated at MPP (same test conditions as above). The gain in electrical energy yield is 11.6%.

10.4.5 Construction, Operation and Measurement of TEPVIS in Africa

In September 1995 a TEPVIS-tank was ordered at a locksmith's shop in Harare (Zimbabwe). The material used for the construction was galvanized steel [21] (see Fig. 10.20).

Fig. 10.20. Tank for reduction of cell temperature and serving as module mount/ foundation, manufactured in Harare (Zimbabwe) by galvanized sheet steel (without internal convection aid).

Two PQ 10/40 multi-crystalline PV modules (by Telefunken, now ASE-Schott/RWE) have been selected out of a series of twelve, according to similar short-circuit currents and open circuit voltages (see Fig. 10.21). To eliminate even small possible measurement errors, the modules were interchanged and retested after each day of measurement. The figures below are showing the average values of the measurements. The Maximum-Power-Point (MPP) was tracked manually by power metering and a variable ohmic load. Irradiance was measured by a BM 5 pyranometer.

---

[21] The use of stainless steeel would have been more adequate in terms of service life of the system, but for the limited duration of the tests (max. 3 months) galvanized steel was considered as to be sufficient.

Optimization 209

Fig. 10.21. Comparative measurement of TEPVIS (in the back) with a standard reference PV module (in the front), both based on PQ 10/40 together with a BM 5 pyranometer during the measurements in 1995 in Zimbabwe.

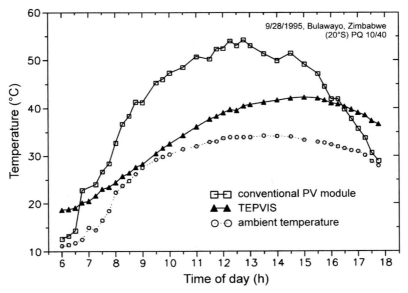

Fig. 10.22. Comparison of module temperatures for a conventional multi-crystalline PV module with TEPVIS (both based on PQ 10/40) during a clear day in Bulawayo, Zimbabwe (20°10' S, 38°37' E).

The results for the tests carried out in Bulawayo, Zimbabwe are presented in the Figures 10.23, 10.24, and 10.25. Despite quite favorable weather conditions -

relatively cold nights and cloudless days (however breezy) - the gains attained at the experiments in Berlin, Germany could not be achieved: The yields increase of TOEPVIS (without optical improvements) over the conventional installation (averaged over two tests with exchanged solar modules) amounted only 6.4%. The reasons for the lower gain could be found in the flatter inclination of construction (20° instead of 33° module elevation angle), and the absence of a "convection accelerator" (a plate of 2/3 of module length, installed inside the tank parallel to the module surface at a distance of 15 cm to it). Both facts reduced thermal convection inside the tank and led to a high thermal stratification (ca. 10 K temperature difference the top and at the bottom of the tank), which caused higher operation temperatures for the module. To avoid the negative effects of stratification, a new design with a tank having the main part of its reservoir above the module is under construction. On the other hand it has to be examined whether the cost reducing effect by substitution of the concrete foundation with the tank is still possible with that new design.

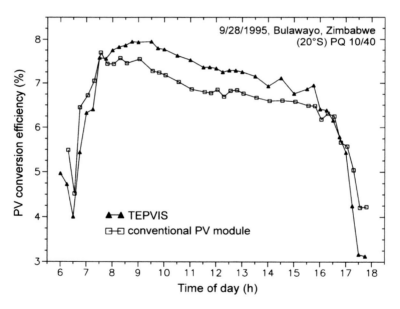

Fig. 10.23. Comparison of PV conversion efficiencies for a conventional multi--crystalline PV module (PQ 10/40) with TEPVIS during a clear day in Bulawayo, Zimbabwe.

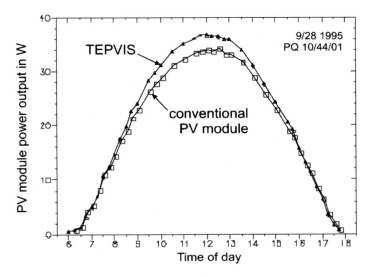

Fig. 10.24. Electrical power output at MPP of TEPVIS during a clear day in comparison to a conventional PV module (Location: Bulawayo, Zimbabwe, 20°10' S, 38°37' E).

10.4.6 The Integrated Solar Home System (I-SHS)

10.4.6.1 Composition of the system

This project has been carried out by Fabian Ochs, a master student of the author, during 2001/02. Figure 10.25 shows the basic layout of the system: The PV generator consists of two parallel-connected, frameless 30 W_p modules. Located in the foundation structure are a maintenance-free lead-acid battery (12 V, 105 Ah) and a 200 W sine inverter (115 V, 60 Hz) with an integrated charge controller (6 A). A water tank cools all components. The output leads to a regular AC plug. All components are contained in a waterproof epoxy fiber glass tank. The prototype is 1.37 m long, 0.76 m high and 0.5 m deep and has a volume of 0.3 m³. It can be transported easily when empty (20 kg), and is fixed when filled up with water at the installation site (320 kg) without necessity to penetrate the roof surface (see Figs. 10.25, 10.26).

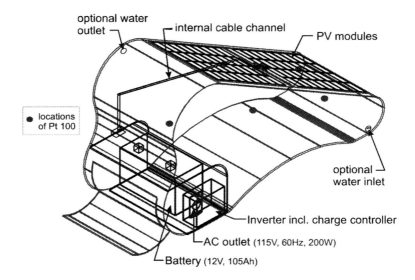

Fig. 10.25. Internal structure of the Integrated Solar Home System (I-SHS).

Fig. 10.26. The Integrated SHS (I-SHS) prototype during tests in Copacabana, Rio de Janeiro, Brazil. Modified design of "Escola de Belas Artes", UFRJ, Rio de Janeiro.

A module elevation angle of 30° was chosen to achieve a good yield even in winter in most parts of Brazil. The tank has a volume of almost 300 liters, which results in a weight of 300 kg, when full. The tank acts as an efficient cooler for the PV modules. The aluminum back of the PV modules allows for good heat transfer to water stored in the tank. The water, with its high thermal capacity, limits solar cell temperatures to a range in proximity to that of ambient temperatures (see Fig. 10.27).

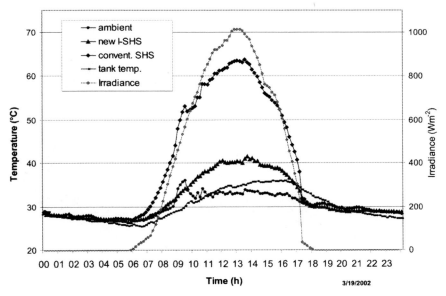

Fig. 10.27. Temperature measurements of the I-SHS during a clear day (see "Irradiance"): The lower module temperature ("new I-SHS") and water temperature in upper part of the container ("tank temp."), in comparison to a conventional SHS and ambient temperature.

The increase of cell temperature relative to ambient temperature was measured for several days in Rio during March 2002 and is shown in Fig. 10.28 as a function of irradiance in comparison to the equivalent values for a conventional SHSs (see Messenger and Ventre 2000, Krauter 1998, Krauter and Schmid 1999).

Despite a relatively wide spread in values, mainly due to wind speed variations, the following linear approximations can be extracted:

$$T_{conv.\ SHS} - T_{ambient} = 0.03 \cdot G\ (W/m^2)^{-1}\ K$$

$$T_{I\text{-}SHS\ upper} - T_{ambient} = 0.012 \cdot G\ (W/m^2)^{-1}\ K$$

$$T_{I\text{-}SHS\ lower} - T_{ambient} = 0.0058 \cdot G\ (W/m^2)^{-1}\ K$$

G stands for global irradiance, T_{SHS} for module operation temperature of a conventional SHS ("Reference Case") as measured Krauter 1998, Krauter and Schmid 1999) or given in literature (Messenger and Ventre 2000, Krauter and Hanitsch 1996). T_{upper} stands for the temperature of the upper module and T_{lower} for the lower module in the I-SHS. All temperatures are given in Kelvin (K) or degree Celsius (°C).

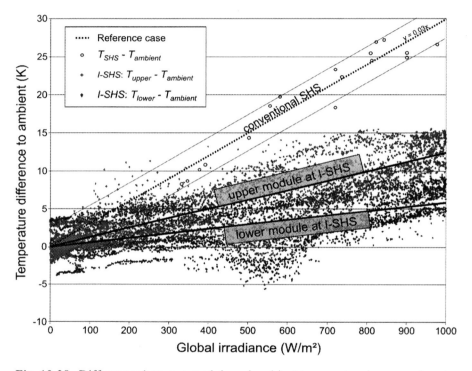

Fig. 10.28. Differences between module and ambient temperature in comparison to the reference case (conventional SHS) plotted as a function of irradiance.

In previous experiments, a reduction in cell temperatures during operating time increases the electrical yield by up to 12% (Krauter 1995, Krauter 1996). Due to the stratification observed, the I-SHS showed just a 9% gain. Forcing circulation in the tank would certainly result in higher electrical yields, on the other hand, stratification serves very well for an optional thermal use of the system. Additionally, the hot water generated is sufficient for the consumption of a small household in Brazil. The upper module can also be replaced by a thermal absorber and would boost hot water generation.

For testing purposes a single prototype was manufactured. Mass production of the I-SHSs (e.g., by recycled PP or PE) would be fast and inexpensive with production costs of less than 50 € for mass production. Material problems related to UV-stability and the drying of the plastic seem to be solved - manufacturers of similar tanks (used as floating docks for boats) give guarantees of ten years. The prototype's form is shaped from a block of expanded polystyrene (EPS). Subsequently the container-tank, consisting of six parts, was laminated using a fiber glass and epoxy resin. To allow for modifications the modules are mounted in a detachable fashion. A cable channel through the tank was installed to simplify maintenance. Additionally, fixed integration of the PV modules within the tank would provide improved performance. Construction time for the prototype was less than a week. Material costs for the prototype have been 420 €.

10.4.6.2 Balance of System Costs (BOS)

Since the foundation, support structure and mounting equipment are no longer required, significant reductions in installation costs and "turn-key" system costs are achieved. Together with improved aspects of maintenance and higher energy yields, PV electricity is becoming more available. Once the I-SHS has been placed at an appropriate site, it has just to be filled with water and is immediately ready to supply power to any AC device from its standard plug. The weight of the tank-container, without inverter and battery, is about seven kg, making transportation easy. When filled with water the container has a weight of more than 300 kg, thus making the system stable enough to withstand any storm without additional fixings. Once placed at an appropriate site the I-SHS is immediately ready to supply small AC loads (lightning, air-fan, radio etc.). Additionally it is capable of supplying the hot water needed for a small household. Several systems can be combined to fulfill higher power needs without a redesign of the system. Without having higher costs than conventional SHSs, and featuring a favorable BOS and the generation of more energy, the I-SHS is an efficient means to successfully electrify remote areas.

10.4.6.3 Benefits of the I-SHS

- Ease of installation
- Significant reduction of system costs
- Increased efficiency via low cell temperature operations
- Increased reliability via pre-manufactured and pre-tested units
- Standard AC output ("Plug and Play")
- Optional use of hot water as a by-product.

10.4.6.4 Further Development

The combination of all suggested improvements leads to a gain in electricity yields of 15–17%. At the same time systems costs could be reduced by 10–15% via pre-assembly and reduced installation costs. Ultimately the PV generation costs could be lowered by 28-38%. Construction can be environmentally sound: Recycling of all materials and components (glass, Si, PE) is entirely possible. Raw materials are available infinitely (glass, Si) or can be made of recycled materials (PE). The product has a long service life and is maintenance free.

Production of a PE structure is about 80 € in mass production, 250 € for small scale production (100 pieces per year) considering initial costs for the deep drawing form or extrusion blow molding 7,000 €. While labor costs are considerably lower in the Third World, manufacturing of the I-SHS there would be very favorable, also in terms of job creation.

11
Summary

The drastic increase of humanity's energy consumption results in an exponential rise in CO_2 emissions related to current predominant types of energy generation technology. The radiation exchange balance between the Earth's surface and space has been altered by a significant increase in CO_2 contents within the Earth's atmosphere as observed over the last decades; now the balance occurs at higher surface temperatures. This is caused by the reduction of optical transmittance of the Earth's atmosphere in the infrared range, at which thermal radiation from the Earth occurs, while the Sun's solar spectrum reaching the ground remains relatively unchanged. Effects of that temperature increase, such as an increase of floods and hurricanes, cause additional damages in the vicinity of 50 billion US$ each year and are reflected in the statistics of Munich Re-insurance already today. On this back ground the need to examine the present energy supply systems concerning their carbondioxide intensity during their life cycle becomes obvious.

This work examines energy- and CO_2 balances of photovoltaic power plants during their live cycles, including production, operation, dismantling and re-use.

Parameters influencing this balance such as raw materials, production and operation conditions, yield , and recycling-quotes are considered:

- Taken into account are the expenditures on the production of PV power plants, considering their specific material composition and conditions of production processes, even that of the raw materials.
- During the operation phase, all parameters having an influence on the electrical yield for the PV power plant such as cell reaching irradiance and operation temperature are modeled and examined in detail. Several suggestions for improvements are given and tested.
- The use of recycled material has a crucial impact on the energy expenditures of manufacturing: The exclusive use of secondary materials results in an energy saving of 92% for aluminum, for copper 73% and for glass 67%. This potential is not just a theoretical one; currently industrial recycling-quotas (in Germany) are already in the vicinity of 31% to 35% (up to 87% in the electrical engineering sector) for aluminum, 48% to 55% for copper and 45% for glass.

If the photovoltaic power plant is getting recycled after dismantling at the end of its usable lifetime, the material related energy-expenditures reduce themselves once more. Initial examinations revealed that the energetic expenditure for solar cells can be lowered by one magnitude if recycled material is used. Due to insufficient data at the manufacturing plants a detailed analysis of this effect could not be carried out yet. The modeling of the CO_2 balance is more complex. For the production of a PV power plant, its components and its raw materials, the carbon dioxide intensity of the country of production has to be taken into account. Already within Europe the corresponding specific CO_2 emissions can differ by a factor of 27.6 (Netherlands at 442 g/kWh vs. Norway at 16 g/kWh). Increased global trade, where locations of production and application of most materials, components and products are rarely in proximity, in conjunction with the tendency to change suppliers frequently, is making a solid, lasting statement on CO_2 balance enigmatic.

Under present basic conditions, an exemplary comparison of Germany with Brazil shows that the highest CO_2-reduction (26,805 kg per kW_p of installed PV power plant) is observed for the production of PV power plants in Brazil and the local substitution of off-grid diesel-generators by PV-power plants based on single-crystalline silicon technology. For production and installation in Germany, power plants based on multi-crystalline solar cells have some better CO_2-reduction abilities: 8,677 kg/kW_p vs. 7,792 kg/kW_p with single-crystalline technology.

Such calculations are significant for carbon trading: under unfavorable side conditions, for example production in Germany (with its relatively high CO_2 intensity of 0.56 kg/kWh) and installation with a grid-connection in Brazil (with its very low CO_2 intensity), application of PV may even cause a negative CO_2 reduction: -1,009 kg/kW_p (worst case).

All data are computed with conservative assumptions for the recycling quote (25%) as well as for the manufacture of raw materials (only local production). The balance can be influenced positively by importing raw materials from countries with a favorable energy-mix (e.g., Norway, Iceland).

To further improve CO_2 reduction by photovoltaic power plants, further measures have been proposed, simulated and tested: Via an accurate yield prediction, considering all relevant optical, thermal and electrical parameters, the planning-expenses can be reduced considerably by minimization of yield uncertainties. Beside the modeling, practical measures, primarily of optical and thermal nature, have been proposed to increase the electrical yield of a PV power plant.

Theoretical model-assumptions could be confirmed by practical measurements in Australia, Brazil, Germany and Zimbabwe. Examples: Irradiance reflection losses could be reduced by better matching the refractive indices, and the application of optical structures will allow to generate about 4% more electricity. Diminution of the operation temperature of the PV-Generators through large heat-capacities (water tank) results in an electrical yield gain of 6% to 12%. Additionally, the substitution of the conventional concrete-foundation by the water tank leads to a reduction of material and labor expenses. For the same yield a smaller generator becomes necessary, which further reduces material and energy requirements for production, and therefore increases the CO_2 reduction potential during the life cycle of PV power plants.

12
Appendix

12.1 List of Symbols and Abbreviations

Abbrev.	Meaning	Unit
a	year	
a-Si	amorphous silicon	
A	area	$[A] = m^2$
A_F	front side area	$[A_F] = m^2$
A_R	back side area	$[A_R] = m^2$
ABS	acrylonitrile-butadiene-styrene (Polymer)	
AC	alternating current	
Albedo	irradiance caused by reflections from ground	
AM	relative air mass	$[AM] = -$
ARC	anti-reflective-coating	
C_p	heat capacity	$[C_p] = kJ\ K^{-1}$
C'_p	heat capacity related to an area	$[C'_p] = kJ\ K^{-1}\ m^{-2}$
c_p	specific heat capacity	$[c_p] = kJ\ kg^{-1}\ K^{-1}$
c-Si	single crystal silicon	
CdS	cadmium-sulfite	
CdTe	cadmium-telluride	
CuInSe$_2$	cooper-indium-di-selenite	
CVD	chemical vapor deposition	
d	thickness of a layer	$[d] = m$
d_i	thickness of layer no. i	$[d_i] = m$
D	length of path passed by irradiance	$[D] = m$

DC	direct current	
E	irradiance (European norm)	$[E]$ = W m^{-2}
E_0	extraterrestrial irradiance	$[E_0]$ = W m^{-2}
E_{diff}	irradiance of the diffuse component	$[E_{diff}]$ = W m^{-2}
E_{dir}	radiance of the direct component	$[E_{dir}]$ = W m^{-2}
E_{cell}	cell-reaching irradiance	$[E_{cell}]$ = W m^{-2}
EG-Si	electronic grade silicon (99.99999% pure)	
ERZ	energy-pay-back-time	
EVA	ethylene-vinyl-acetate (co-polymer)	
FF	form-factor of the I-V characteristics	$[FF]$ = -
FZ	floating zone	
g	acceleration by Earth's gravity (9.8067 m s^{-2})	$[g]$= m s^{-2}
G	irradiance (American norm)	$[G]$ = W m^{-2}
GaAs	gallium-arsenide	
Ge	germanium	
Gr	Grashof number	$[Gr]$=-
h	Planck's constant of action (6.626 · 10^{-34} J s)	$[h]$ = N m s
h	heat transfer coefficient	$[h]$ = W m^{-2} K^{-1}
h	height (above sea level)	$[h]$ = m
HR	relative humidity	$[HR]$ = %
H_2	molecular hydrogen	
He	helium	
I_D	conducting-state current (forward current) ($I_D \approx 0.1\,I_{max}$)	$[I_D]$ = A
I_{photo}	photoinduced current	$[I_{photo}]$ = A
I_{mp}	current at maximum power point	$[I_{mp}]$ = A
I_{sc}	short circuit current	$[I_{sc}]$ = A
IR	infrared radiation (λ > 800 nm)	
k	thermal conductivity	$[k]$ = W m^{-1} K^{-1}
k	Boltzmann constant (1.3807 · 10^{-23} J K^{-1})	$[k]$ = N m K^{-1}
k	imaginary part of the optical refractive index, coefficient of extinction	$[k]$ = -

KEA	accumulated energy requirements (AEE)	[KEA] = kWh
KEV	accumulated energy consumption (AEC)	[KEV] = kWh
KOH	caustic potash solution	
l	liter (1 l = 1 dm^3)	
l_{ch}	characteristic length	$[l_{ch}]$ = m
L	length of module	$[L]$ = m
LCA	life cycle analysis	
MG-Si	metallurgic grade silicon	
MP	maximum power	
MPP	maximum power point	
MPPT	maximum power point tracking	
m-Si	mono- or single-crystalline silicon	
MWS	multi-wire-slurry (method for sawing wafers)	
n	optical refractive index	$[n]$ = -
$\hat{n}_i$	complex refractive index of layer i	$[\hat{n}_i]$ = -
n_1	optical refractive index of the original media	$[n_1]$ = -
n_2	optical refractive index of the entrance media	$[n_2]$ = -
N	number of day of a year	$[N]$ = -
NEV	non-energetic consumption	
NOC	nominal-operating-conditions (at E = 800 W/m², w = 1 m/s, $\vartheta \triangleq$ NOCT)	
NOCT	nominal operating cell temperature	[NOCT] = °C
Nu	Nusselt number	$[Nu]$ = -
OC	open circuit	
p	air pressure	$[p]$ = N m^{-2}
p_0	nominal pressure (1.013 · 10^6 N m^{-2})	$[p_0]$ = N m^{-2}
poly-Si	poly-(multi)-crystalline silicon	
P	power, heat flow	$[P]$ = W
P_{el}	electrical power	$[P_{el}]$ = W
P_{in}	irradiance power into cell	$[P_{in}]$ = W
P_{mp}	power of a solar cell for optimal load adaptation	$[P_{mp}]$ = W
P_n	nominal power (type plate power at STC)	$[P_n]$ = W$_p$

Pr	Prandtl's number	$[Pr] = 1$
PC	poly-carbonate	
PE	poly-ethylene	
PMMA	acrylic glass (Plexiglas®)	
PTFE	poly-tetra-fluor-ethylene (Teflon®)	
PP	ploy-propylene	
PV	photovoltaic	
PVC	poly-vinyl-chloride	
PVF	poly-vinyl-fluoride (Tedlar®)	
q	charge of an electron ($1.6022 \cdot 10^{-19}$ A s)	$[q]$ = A s
Q	quantity of heat	$[Q]$ = W s
Q	heat flow	$[Q]$ = W
Q_{fc}	heat flow by forced convection	$[Q_{fc}]$ = W
Q_{nc}	heat flow by natural convection	$[Q_{nc}]$ = W
Q_{rad}	heat flow by radiation	$[Q_{rad}]$ = W
Q_F	heat flow through the front side of the module	$[Q_F]$ = W
Q_B	heat flow through the back of the module	$[Q_B]$ = W
r	reflection coefficient	$[r]$ = -
$r_\parallel$	reflection coeff. of parallel polarized irradiance	$[r_\parallel]$ = -
$r_\perp$	reflection coeff. of orthogonal polarized irrad.	$[r_\perp]$ = -
R	electrical resistance	$[R] = \Omega$
R	reflectivity	$[R]$ = -
R_0	reflectivity for irrad. at perpendicular incidence	$[R_0]$ = -
R_{xy}	reflectance of a single boundary between material x and material y	$[R_{xy}]$ = -
$R_\parallel$	reflectivity for parallel polarized irradiance	$[R_\parallel]$ = -
$R_\perp$	reflexivity for perpendicular polarized irradiance	$[R_\perp]$ = -
Ra	Rayleigh's number ($Ra = Gr \cdot Pr$)	$[Ra]$ = -
RC	real/reported Conditions	
Re	Reynold's number	$[Re]$ = -
SG-Si	solar-grade silicon	
SC	short circuit	

Appendix

Si	silicon	
STC	Standard-Test-Conditions ($\vartheta_c = 25\,°C$; $E = 1000\ W\ m^{-2}$; AM 1.5; perpendicular incidence)	
SRC	standard-real/reported-conditions	
t	time of day in hours	$[t] = h$
T	transmittance	$[T] = -$
T_{xy}	transmittance of a single boundary between material x and material y	
$T_\parallel$	transmittance of parallel polarized irradiance	$[T_\parallel] = -$
$T_\perp$	transmittance of perpendicular polarized irrad.	$[T_\perp] = -$
T	temperature	$[T] = K$
T_A	ambient temperature	$[T_A] = K$
T_C	cell temperature	$[T_C] = K$
T_F	surface temperature at module front side	$[T_F] = K$
T_g	"glass" temperature of Polymers	$[T_g] = K$
T_H	sky temperature	$[T_H] = K$
T_{Hl}	semiconductor temperature (equivalent to T_C)	$[T_{Hl}] = K$
T_i	temperature of layer i	$[T_i] = K$
T_M	module surface temperature (in general)	$[T_M] = K$
T_R	surface temperature at the module backside	$[T_R] = K$
T_s	radiation temperature of a black body	$[T_s] = K$
T_S	melting temperature (melting point)	$[T_S] = K$
TC	temperature coefficient (standardized)	$[TC] = K^{-1}$
TiO_2	titan-dioxide (as anti-reflective-coating)	
u	normalized voltage ($u = U\,q\,(k\,T)^{-1}$)	$[u] = -$
U_D	on-state (forward/conducting) voltage (at $I_D \approx 0.1\,I_{max}$)	$[U_D] = V$
U_{mp}	voltage at the point of maximal power output	$[U_{mp}] = V$
U_{oc}	open circuit voltage	$[U_{oc}] = V$
u_{oc}	normalized open circuit voltage	$[u_{oc}] = -$
$U_{oc\,25°C}$	open circuit voltage at cell temp. of 25 °C	$[U_{oc\,25°C}] = V$
U_{String}	voltage of a sting (cells in series connection)	$[U_{String}] = V$
U_{Vl}	voltage loss by partial shading	$[U_{Vl}] = V$

U_{V2}	voltage loss by reverse current blocking diode	$[U_{V2}]$ = V
V	degree of polarization	[V] = -
w	wind speed	[w] = m s^{-1}
W_p	PV power output in Watt under STC	
WR	inverter transforming DC to AC	
W	(primary-) energy	[W] = J
ZnO	zinc oxide (used as a ARC)	
ZnS	zinc sulfate (used as a ARC)	
ZrO_2	zircon dioxide (uses as a ARC)	
α	absorption coefficient	$[\alpha]$ = m^{-1}
α	heat transfer coefficient (European norm)	$[\alpha]$ = W K^{-1} m^{-2}
α_{fc}	heat transfer coefficient, forced convection	$[\alpha_{eK}]$ = W K^{-1} m^{-2}
α_{nc}	heat transfer coefficient, natural convection	$[\alpha_{fK}]$ = W K^{-1} m^{-2}
α_{Fconv}	convective (fc+nc) heat transfer coeff. of the module front	$[\alpha_{Fkonv}]$ = W K^{-1} m^{-2}
α_{Bconv}	convective (fc+nc) heat transfer coeff. of the module back	$[\alpha_{Rkonv}]$ = W K^{-1} m^{-2}
α	azimuth angle (deviation from North)	$[\alpha]$ = °
α_M	azimuthal orientation of the module (front)	$[\alpha_M]$ = °
α_P	azimuth angle of a point P at the sky sphere	$[\alpha_P]$ = °
α_S	azimuth angle of the sun	$[\alpha_S]$ = °
β	thermal volumetric expansion coefficient	$[\beta]$ = K^{-1}
γ_M	elevation angle of the module	$[\gamma_M]$ = °
γ_S	elevation angle of the sun	$[\gamma_S]$ = °
$\Delta\gamma_S$	additional elevation angle of the sun (caused by refraction of the atmosphere)	$[\Delta\gamma_S]$ = °
δ_S	angle covering the Sun's diameter ("Sun angle" = 0.54°)	$[\delta_S]$ = °
ε	Emittance of a surface (in Infrared)	$[\varepsilon]$ = -
ε_λ	Emittance at a wavelength λ	$[\varepsilon_\lambda]$ = -
ϵ_0	Permittivity constant ϵ_0 = 8.85 10^{-12} F m^{-1}	$[\epsilon_0]$ = F m^{-1}
ϵ	$\epsilon = \epsilon_1 - i\epsilon_2$ dielectric constant	$[\varepsilon]$ = F m^{-1}

Symbol	Description	Units
η	efficiency	$[\eta] = -$
η_{PV}	photovoltaic conversion efficiency	$[\eta_{PV}] = -$
η_{DZ}	dynamic viscosity of fluid	$[\eta_{DZ}] = \text{N s m}^{-2}$
θ_{in}	angle of incidence (referenced to normal of interface)	$[\theta_{in}] = °$
θ_{out}	angle of refraction	$[\theta_{out}] = °$
θ_i	angle of incidence at layer i	$[\theta_i] = °$
θ_p	Brewster's- (or polarization-) angle	$[\theta_p] = °$
ϑ_A	ambient temperature	$[\vartheta_A] = °C$
$\vartheta_A(t)$	course of ambient temperature in time	$[\vartheta_A(t)] = °C$
$\hat{\vartheta}_A$	peak of ambient temperature	$[\hat{\vartheta}_A] = °C$
$\bar{\vartheta}_A$	average ambient temperature	$[\bar{\vartheta}_A] = °C$
ϑ_C	cell temperature	$[\vartheta_C] = °C$
λ	wavelength of radiation	$[\lambda] = \text{nm}$
λ	thermal conductivity	$[\lambda] = \text{W K}^{-1} \text{m}^{-1}$
ρ	density	$[\rho] = \text{kg m}^{-3}$
ρ_L	density of air	$[\rho_L] = \text{kg m}^{-3}$
ρ_0	density at seal level	$[\rho_0] = \text{kg m}^{-3}$
ρ	reflexivity	$[\rho] = -$
ρ_i	reflexivity of layer i	$[\rho_i] = -$
$\bar{\rho}_i$	internal reflexivity of layer i	$[\bar{\rho}_i] = -$
ρ_{nm}	reflexivity of the system consisting of layers n and m	$[\rho_{nm}] = -$
ρ_i	reflectance of whole slab i	$[\rho_i] = -$
ρ_{ijk}	reflectance of an optical three slab system consisting of slab i, j and k	$[\rho_{ijk}] = -$
$\bar{\rho}_{ik}$	inner reflectance of an optical two slab system consisting of slab i and k	$[\bar{\rho}_{ik}] = -$
ρ_{12}	view factor	$[\rho_{12}] = -$
σ	Stefan-Boltzmann's constant ($5.670 \cdot 10^{-8} \text{ W m}^{-2} \text{ K}^{-4}$)	$[\sigma] = \text{W m}^{-2} \text{ K}^{-4}$
τ	Transmittance	$[\tau] = -$
τ_i	Transmittance of a single layer i	$[\tau_i] = -$

τ_{123}	Transmittance of an entire optical system consisting of layers 1, 2 and 3	$[\tau_{123}] = -$
$\bar{\tau}_i$	internal transmittance of layer i	$[\bar{\tau}_i] = -$
$\tau_{ij,1}$	transmittance of the first ray of an optical two slab system consisting of slab i and j	$[\tau_{ij,1}] = -$
τ_{ijk}	transmittance of an optical three slab system consisting of slab i, j and k	$[\tau_{ijk}] = -$
$\bar{\tau}_{ik}$	inner transmittance of an optical two slab system consisting of slab i and k	$[\bar{\tau}_{ik}] = -$
Ø	average of value	
$\parallel$	electrical field vector parallel to the plane of incidence	
$\perp$	electrical field vector perpendicular to the plane of incidence	
$[x]$	means "unit of x". Opposite to that at approximation equations in which just the numerical values of the variable x are needed, this is indicated by "x in units of y" in the explanation.	

12.2 Tables

Table A1. Conversion of different energy units

Energy unit	Btu	J (Ws)	kWh	cal	t SKE (tons of coal equivalent)	TOE (tons of oil equivalent)
1 Btu =	1	1,055	$0.2930 \cdot 10^{-3}$	252	$35.9 \cdot 10^{-9}$	$25.13 \cdot 10^{-9}$
1 J =	$9.481 \cdot 10^{-3}$	1	$277.8 \cdot 10^{-9}$	2388	$34.1 \cdot 10^{-12}$	$23.87 \cdot 10^{-12}$
1 kWh =	3413	$3.600 \cdot 10^{6}$	1	$8.601 \cdot 10^{5}$	$0.1228 \cdot 10^{-3}$	$8.596 \cdot 10^{-4}$
1 cal =	$3.968 \cdot 10^{-3}$	4.186	$1.163 \cdot 10^{-6}$	1	$0.1429 \cdot 10^{-9}$	$0.1000 \cdot 10^{-9}$
1t SKE =	$27.78 \cdot 10^{6}$	$29.31 \cdot 10^{9}$	8141	$7.00 \cdot 10^{9}$	1	0.7
1 TOE =	$39.69 \cdot 10^{6}$	$41.87 \cdot 10^{9}$	11630	$10.0 \cdot 10^{9}$	1.4286	1

Conversion of larger SI energy units:
1 EJ=1,000 PJ=1,000,000 TJ=1,000,000,000 GJ=10^{9} GJ =10^{12} MJ =10^{15} kJ =10^{18} J
Energy flux: 1 W/m^2 = 0.317 Btu/(h ft^2), 1 langley (ly) = 1 cal/cm^2 = 3.687 btu/ft^2

Table A2. Calorific value and CO_2 intensity of fossil fuels

Type of fuel	Density in kg/m³	Calorific value in MJ/kg	Reference	CO_2 in kg/MJ	Reference
Straw		16	Kleemann 1993		
Waste paper	700–1200	17	Kleemann 1993		
Sugar cane		15	Kleemann 1993		
Wood (general)	400–800	13.3	Gieck 2005		
		14.65	BMWi 1996		
Ash	750	18.6	Kleemann 1993		
Beech	720	18.8	Kleemann 1993		
Oak	850	18.3	Kleemann 1993		
Confiers (e.g. spruce)	480–620	19.0	Kleemann 1993		
Peat	190	12.0	Mende 1981		
Fuel peat	500	14.24	BMWi 1996		

Type of fuel	Density in kg/m³	Calorific value in MJ/kg	Reference	CO_2 in kg/MJ	Reference
Lignite	1200–1400	9.60	Gieck 2005	0.111	Kolb 1989
		8.48	BMWi 1996	0.111	Lewin 1993
Lignite bricks	1250	20.0	Mende 1981		
		19.47	BMWi 1996		
Lignite coke	1200–1500	29.936	BMWi 1996		
Coal	1350	29.761	BMWi 1996	0.0917	Lewin 1993
Furnace coke	1100–1400	30.1	Gieck 2005		Faber 1996
Anthracite	1300–1500	33.4	Gieck 2005	0.0751	
Raw mineral oil	730–940	42.8	BMWi 1996		
Diesel	830	42.1	Gieck 2005	0.0777	Kolb 1989
Fuel oil EL (extra light)	860	41.843	Gieck 2005 Kleemann 1993	0.0621	Faber 1996
Fuel oil L (light)	1100	42.73	BMWi 1996	0.0777	Lewin 1993
Fuel oil H (heavy)	> 1200	40.61	BMWi 1996	0.0833	Lewin 1993
Petroleum	810	42.00	Mende 1981		
		43.00	BMWi 1996		
Kerosine	720	43.54	BMWi 1996		
Gasoline	780	42.50	Gieck 2005		
Ethanol	789	26.90	Gieck 1989		
Methanol	792	20.00	Mende 1981		
Liquid natural gas		45.99	BMWi 1996		

Type of fuel	Density in kg/m^3	Calorific value in MJ/kg	Reference	CO_2 in kg/MJ	Reference
Mineral oil gas		40.3 MJ/m^3	BMWi 1996		
Natural gas		31.74 MJ/m^3	BMWi 1996	0.0528	Lewin 1993
Town gas	0.580	15.99 MJ/m^3	BMWi 1996	0.0461	Faber 1996
Butane	2.680	124 MJ/m^3	Mende 1981		
Propane	2.010	94 MJ/m^3	Mende 1981		
Methane	0.720	36 MJ/m^3	Mende 1981		
Sewage gas		16 MJ/m^3	BMWi 1996		
Carbon-monoxide	1.250	13 MJ/m^3	Mende 1981		
Hydrogen	0.090	119.90	Gieck 2005		
Electricity (Germany): Primary energy balance		9.37 MJ/kWh	BMWi 1996	0.53 (mix of german power gen.)	
Final energy consumption		3.60 MJ/kWh	BMWi 1996		
1 kg Carbon				3.87 kg CO_2	Shell 1997

Table A3. CO_2 emissions for the generation of 1 kWh of electrical energy

Type	Fuels in kg/kWh	Transport preparat. in kg/kWh	Construct. of power plant 1 in kg/kWh	Sum CO_2 in kg/kWh	Sum CO_2 equiv. in kg/kWh	Reference
Lignite					1.244	Lewin 1993b
				1.1356	1.1466	DB 1995
	0.99					Kaltschmitt 1995b
Coal					1.206	Lewin 1993b
	0.781	0.0495	0.0044-0.0072	0.834-0.835		Stelzer 1994
	0.781			0.834-0.917		Kaltschmitt 1995
				0.9177	1.0499	DB 1995
	0.781	0.0001		0.844	0.9384	Kaltschmitt 1997
Natural gas	0.365			0.373-0.416		Kaltschmitt 1995
	0.361	0.006		0.4235	0.4501	Kaltschmitt 1997
Combin. Gas				0.4076	0.4387	DB 1995
					0.453	Lewin 1993b
Diesel-co-gen.				0.3265	0.3407	DB 1995
Wind, general	0				0.011	Lewin 1993b
Wind power 4.5 m/s	0			0.0163		Stelzer 1994
	0			0.021-0.036	0.024-0.040	Kaltschmitt 1997
Wind power 5.5 m/s	0			0.0108		Stelzer 1994
	0		0.013-0.022	0.013-0.022		Kaltschmitt 1995
	0			0.015-0.023	0.016-0.025	Kaltschmitt 1997
	0				0.042-0.046	Kaltschmitt 2003

Appendix 233

Type	Fuels in kg/kWh	Transport preparat. in kg/kWh	Construct. of power plant 1 in kg/kWh	Sum CO_2 in kg/kWh	Sum CO_2 equiv. in kg/kWh	Reference
Wind power 6.5 m/s	0			0.0081		Stelzer 1994
	0			0.011-0.018	0.012-0.020	Kaltschmitt 1997
	0				0.0295-0.0339	Kaltschmitt 2003
Wind power 7.5 m/s	0				0.0226-0.0257	Kaltschmitt 2003
Hydro power	0			0.0072		Stelzer 1994
	0		0.007-0.020	0.007-0.020		Kaltschmitt 1995
Small hydro	0			0.014-0.015	0.015-0.017	Kaltschmitt 1997
	0				0.0165-0.0211	Kaltschmitt 2003
Large hydro	0			0.006-0.010	0.007-0.011	Kaltschmitt 1997
	0				0.0100-0.0164	Kaltschmitt 2003
PV (general)	0		0.170-0.260	0.170-0.260		Kaltschmitt 1995 (Central Europe)
	0				0.228	Lewin 1993b
PV (sc-Si)	0			0.150-0.250		Hagedorn 1990
	0			0.0994		Wagner et al. 1993
	0			0.206-0.318		Voß 1993
	0			0.247-0.318		Brauch 1997
	0			0.188	0.200-0.202	Kaltschmitt 1997
	0				0.238-0.279	Kaltschmitt 2003
	0				0.041	Alsema 2005

Type	Fuels in kg/kWh	Transport preparat. in kg/kWh	Construct. of power plant 1 in kg/kWh	Sum CO_2 in kg/kWh	Sum CO_2 equiv. in kg/kWh	Reference
PV (mc-Si)	0			0.110-0.250		Hagedorn 1990
	0			0.232-0.298		Stelzer 1994 Brauch 1997
	0			0.100-0.170	0.115-0.185	Staiß 1995
	0			0.316-0.318	0.336-0.339	Kaltschmitt 1997
	0				0.199-0.235	Kaltschmitt 2003
					0.031	Alsema 2005
PV (rib-Si)					0.026	Alsema 2005
PV (a-Si)	0			0.100-0.170		Hagedorn 1990
	0			0.206-0.265		Brauch 1997
	0			0.141-0.163	0.150-0.176	Kaltschmitt 1997
	0				0.123-0.130	Kaltschmitt 2003
PV (CdTe)	0			0.0517		Gemis 1992
	0		517	0.0517	0.0612	DB 1995
PV (CIS)	0			0.015		Zittel 1992
PV (CdTe, CIS)	0			0.03	0.0325	Staiß 1995
All power plants in Germany				0.530		Bloss et al. 1992

At the category "Transport and preparation" miner's work are included for coal and pipelines are included for natural gas.
Wind speeds are related to yearly averages and are measured 50 meters above ground for Kaltschmitt 2003 and 10 meters above ground for other references.
PV for Kaltschmitt 1995 and 2003 is given for central European conditions.
For the calculation of CO_2 equivalents the emitted quantities of CH_4 and N_2O are converted into their molar greenhouse effect potential according to Table 2 (Mol weights: H 1g/mol; C 12 g/mol; N 14 g/mol; O 16 g/mol; F 19 g/mol; for CO_2 44g/mol; CO 28 g/mol; O_3 48 g/mol; CH_4 16 g/mol; N_2O 44 g/mol) Calculations Staiß 1995: mc-Si: CH_4: 0.000432 kg/kWh; N_2O: 0.0000072 kg/kWh; CdTe, CIS: CH_4: 0.000072 kg/Wh; N_2O: 0.0000012 kg/kWh.

Table A4. Energy intensities in different economical sectors: Energy requirement per monetary value generated (Ref.: Spreng 1995, resp. Schulze et al. 1992)

Type of Economical Sector	MJ / € primary energy	Type of Economical Sector	MJ / € primary energy
Products, Materials, Components:		Electrical engineering	9.4
Agriculture	189.7	Precision mechanics, Optics	8.2
Forestry, fishery	42.3	Iron, sheet, metal proc. ind.	15.5
Water	21.3	Music & sports equipment, toys, jewelry	10.8
Mining (without coal, oil, natural gas)	31.5	Wood	20.5
Chemistry	33.1	Wood products	10.1
Plastics	16.6	Cellulose, pulp, paper, carton	40.9
Rubber	13.7	Paper- and carton products	19.2
Building material, quarrying	23.5	Printing office, duplication	13.9
Fine ceramics	19.0	Leather, leather products	9.8
Glass and glass products	25.6	Textiles	17.0
Iron and Steel (without blast-furnace gas)	56.3	Clothing	9.0
Non-iron-metals	45.6	Food (without drinks)	16.4
Foundry products	20.5	Drinks	12.3
Drawing industry, rolling mill	22.1	Tobacco products	2.7
steel and light alloy production.	13.1	Civil engineering	11.1
Mechanical engineering	9.6	Completion	8.4
Office machinery, office equip.	10.6	*Services:*	
		Wholesale business	4.9
Street vehicles	11.9	Retail business	7.6
Naval vehicles	13.3	Railways	32.2
Air- and space-vehicles	7.2	Navig., waterways, harbors	29.5

Type of Economical Sector	MJ / € primary energy
Mail, telecom	5.9
Other transport	13.7
Banking	19.8
Insurances	4.3
Renting of buildings, lodging	6.5
Hotel and restaurant industry	10.2
Science, art, publishing	9.2
Health and veterinary	5.5
Other services	4.9
Local authorities	7.2
Social security	8.0
Services w/o rewards (house keeping etc.)	5.9

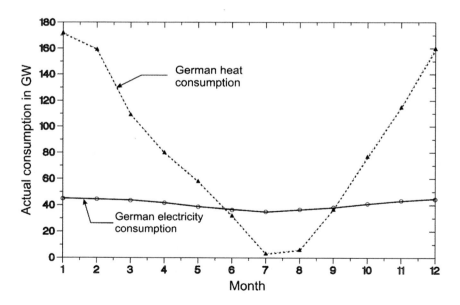

Fig. A1. Seasonal power consumption profile of Western Germany: Electricity and heat load during a typical year (Reference: Heinloth 1996).

Table A5. Energy requirements for the production of different materials

Material to be produced	Primary energy requirem. in MJ/kg	Reference	CO_2 emissions in kg/kg	CO_2 equiv. in kg/kg	Reference
Steel	24.0	Spreng 1995	2.047		Lewin 1993a,b
	30.0	Kaltschmitt 1995	3.00		Kaltschmitt 1995
	20.0 [1]	Kaltschmitt 1997	1.69 [1]	1.85 [1]	Kaltschmitt 1997
Steel (40% recycled)	23.65	Kaltschmitt 2003		1.91	Kaltschmitt 2003
Oxygen steel (primary)	16–27	Hütte 2004			
Steel sheet	25.20	DB 1995			
Electro steel (secondary)	10–18	Hütte 2004			
Cooper	95.0	Kaltschmitt 1995	8.800		Kaltschmitt 1995
		Fritsche 1989	4.901		Lewin 1993a,b
	98.0	Mauch 1995	7.30		Mauch 1995
	88.6 [2]	Kaltschmitt 1997	5.08 [2]	5.36 [2]	Kaltschmitt 1997
Cooper (recycling rate 56%)	37.08	Kaltschmitt 2003		2.80	Kaltschmitt 2003
Cooper tubes	49.0	Wagner 1995a			
Aluminum	250.0	Kaltschmitt 1995	25.00		Kaltschmitt 1995
	250.0	Spreng 1995	12.968		Lewin 1993a,b
	245.0	Mauch 1995	15.10		Mauch 1995
Primary aluminum (based on bauxite)	160–240	Hütte 2004		23.1	Lenzen 1998
	250	Lenzen 1998			
Aluminum plate	260.20	Wagner 1995			
	252.00	DB 1995			
Aluminium (63% recycled)	113.9 [3]	Kaltschmitt 1997	6.70 [3]	7.03 [3]	Kaltschmitt 1997
	107.94	Kaltschmitt 2003		7.57	Kaltschmitt 2003

Material to be produced	Primary energy requirem. in MJ/kg	Reference	CO_2 emissions in kg/kg	CO_2 equiv. in kg/kg	Reference
Secondary aluminum (from scrap)	12.0-20.0	Hütte 2004			
Lead (from ore)	51.0	Lenzen 1998		4.70	Lenzen 1998
Concrete			0.285		Lewin 1993
	0.490	Kaltschmitt 2003		0.130	Kaltschmitt 2003
Concrete (PZ)	0.210 1.415	Hantsche 1993	0.144		Hantsche 1993
Concrete (HOZ)	0.210 0.574	Hantsche 1993	0.0344		Hantsche 1993
Cement	4.0	Kaltschmitt 1995	0.94	0.96	Kaltschmitt 1995
	3.6	DB 1995	0.855		Lewin 1993a,b
	4.4	Mauch 1995	0.40		Mauch 1995
	1.18	Kaltschmitt 1997	0.94	0.96	Kaltschmitt 1997
Portland cement	0.106 7.930	Hantsche 1993	0.994	0.994	Hantsche 1993
Cement (blast furnace)	0.106 2.100	Hantsche 1993	0.205		Hantsche 1993
Glass	20.0	Kaltschmitt 1995	1.90		Kaltschmitt 1995
	10.0	Spreng 1995	0.49		PEC 1996
Float glass	14.0	Wagner 1995b	0.935		Lewin 1993a,b
	8.2	Hagedorn 1996	0.59		Hagedorn 1996
	6.98 14.80	Hantsche 1993	0.539		Hantsche 1993
Float glass (electr. add. heating)	5.23 14.00	Hantsche 1993	0.741		Hantsche 1993
Glass bottles (new)	10.8	DB 1995			DB 1995
Glass bottles (50% recycled)	7.2	DB 1995			DB 1995
Paper (bleached)	79.2	DB 1995			DB 1995

Material to be produced	Primary energy requirem. in MJ/kg	Reference	CO_2 emissions in kg/kg	CO_2 equiv. in kg/kg	Reference
Paper (100% recycled)	18.0	DB 1995			DB 1995
Polystyrene (EPS)	82.0	Spreng 1995			Spreng 1995
	75.3	Wagner 1995a			
	75.0	Hütte 2004			
Polyurethane (PUR)	90.7	Wagner 1995a			
	95.0	Spreng 1995			Spreng 1995
	190.0	Lenzen 1998		15.6	Lenzen 1998
Polyethylene (PE-HD, PE-LD)	80.0	Spreng 1995			
	73.0	Wagner 1995a			
	68.0	Hütte 2004	1.79		PEC 1996
PVC	48.0	Hütte 2004			
PVC (from crude oil)	74.0	Lenzen 1998	7.60		Lenzen 1998
Plastics (in general)	65.0	Kaltschmitt 1995	6.00		Kaltschmitt 1995
	72.0	DB 1995	1.964		Lewin 1993a,b

[1] 40% recycled components, 80% oxygen steel, 20% electric steel
[2] 40% recycled components
[3] 50% recycled components

Table A6. Coal mining subsidies in Germany (Stromthemen 4/97, VDEW)

Year	1997	1998	2005
Subsidies in billion €	4.54	4.72	2.81
Number of jobs	84,000		36,000
Subsidy per year per job in €	54,048		77,947
Subsidy per kWh generated in €	0.051		0.063

cost of German coal: 183.67 €/t, subsidy: 142.86 €/t (adaptation to the price at the world market 40.82 €/t), thermal energy of coal: 29,761 MJ/t = 8,287 kWh/t, electr. energy generation: 2,756 kWh/t (efficiency for electricity gen.: 0.33)

Table A7. Composition of electrical energy generation in Germany 2004 (References: VDEW 2005, DEWI 2005)

Type of energy	Net- generation in billion kWh	Participation in %
Nuclear Power	158.4	27.8
Lignite	146.0	25.6
Coal	127.1	22.3
Natural gas	59.2	10.4
Liquid fuels (diesel and others)	9.2	1.6
Hydro power	20.0	3.7
Wind power	25.0	4.4
PV	0.5	0.1
Waste	2.1	0.4
Biomass	5.2	0.9
Others	16.5	2.9
Sum	570.1	100

Total installed capacity in Germany in 2004: PV 708 MW_p; Wind 16,629 MW

Latest data (BMU, February 2006) indicates that in 2005 renewable energies shared 10.2% of Germany's electricity consumption (wind: 4.3%, hydro: 3.5%, PV: 0.2%, biomass & waste: 2.2%), and decreased CO_2 emissions by 83 million metric tons.

Appendix 241

Table A8. Energy requirements and CO_2 emissions for transportation of goods

Type of transport	Energy direct in MJ/(t km)	CO_2 direct in kg/(t km)	direct plus grey energy in MJ/(t km)	CO_2 total in kg/(t km)	Type of fuel	References for energy data (1996 with CO_2)
Container freighter (overseas) 47,000 t 43 km/h	0.09	0.0075	0.12	0.0094	100% HO	Frischknecht 1994
	0.029	0.00692	0.12	0.0086	100% HO	Frischknecht 1996
Freighter (inland navigation)	0.48	0.04	0.84	0.063	100% D	Frischknecht 1994
	0.51	0.0365	0.844	0.05863	100% D	Frischknecht 1996
Rail	0.48	0.556	1.39	0.1137	51% E 49% D [1]	Frischknecht 1994
	0.319	0.104	0.956	0.0513	80% E 20% D	Frischknecht 1996
	0.9			0.08		Lenzen 1988
Truck (40 t) 50% load	1.16	0.0967	2.16	0.1606	100% D	Frischknecht 1994
	1.00	0.079	2.08	0.13637	100% D	Frischknecht 1996
Truck (28 t) 40% load	1.87	0.156	3.15	0.2378	100% D	Frischknecht 1994
	1.80	0.14	3.14	0.20716	100% D	Frischknecht 1996
Truck (16 t) 40% load	2.78	0.232	4.62	0.3496	100% D	Frischknecht 1994
	3.40	0.259	5.20	0.3458	100% D	Frischknecht 1996
Delivery van (< 3.5 t) 30% load	6.03	0.503	9.99	0.756	30% Gu 50% Gl 20% D	Frischknecht 1994
	14.1	1.03	24.40	1.5438	33% Gu 38% Gl 29% D	Frischknecht 1996
Car [2] average 1.4 passeng. = 140 kg	27.9	2.325	47.60	35.836	60% Gl 20% Gu 20% D	Frischknecht 1994
	22.86	1.479	34.00	22.086	60% G 20% Gc 20% D	Frischknecht 1996

Type of transport	Energy direct in MJ/(t km)	CO_2 direct in kg/(t km)	direct plus grey energy in MJ/(t km)	CO_2 total in kg/(t km)	Type of fuel	References for energy data (1996 with CO_2)
Flight (air cargo, long distance)	10.7 13.8 65.1	0.772 ca. 1.07	15.67	ca. 2.3 [3)] 5.40	100% K	Lufthansa 1996 Heinloth 1996 Lenzen 1998

[1)] inclusive maneuvering
[2)] for Western Europe, data by Frischknecht 1994 for 100 kg/1.4 passengers
[3)] own estimation (material of airplane 100% Aluminum)
following CO_2 characteristics have been used for data from Frischknecht 1994:
Electricity 0.53 kg CO_2/kWh (Germany), Fuels 0.3 kg CO_2/kWh, Non-energetic-consumption (grey energy): 0.23 kg CO_2/kWh (Wagner 1996)
For a 18 ton articulated lorry (40 tons with trailer) Mauch 1993 gives an accumulated energy consumption for production (AEC_P) of 399.3 GJ, within its lifetime of 600,000 km, consumption is 14,400 GJ of diesel fuel and 497 GJ of operational media (oil, rubber, steel) (AEE_O).

G gasoline
Gl gasoline leaded
Gu gasoline unleaded
Gc gasoline with catalyst
K Kerosene
D Diesel fuel
HO heavy oil
E electrically powered

Table A9a. Deviation of magnetic North to true North

Location (longitude, latitude)	Magnetic deviation from true North	Yearly change of deviation
Berlin (13°22' E, 52°30' N)	+ 1° 38'	+6.5' per year
Harare (31°1' E, 17°42' S)	- 7° 52'	+1.5' per year
Los Angles (118°14' W, 34°4' N)	+13° 38'	-1.8' per year
Managua (86°20' W, 12°5' N)	+ 1° 10'	-8.3' per year
Rio de Janeiro (43°12' W, 22°54' S)	+21° 17'	-6.2' per year
Peking (116°20' E, 39°54' N)	- 5° 53'	-1.2' per year
Sydney (151°12' E, 33°52' S)	+18° 45'	+3.1' per year

True bearing = Magnetic bearing + Magnetic declination. Data based on IGRF 2000 for June 2001, provided by the Intl. Association of Geo-sciences and Aeronomy (IAGA), Division V, Workgroup 81. Literature: Campbell 1997.

Table A9b. Material requirements of PV power plants

Material	Specific material consumption in t/MW$_p$ after Staiß 1995		Specific material consumption in t/MW$_p$ after Lewin 1993		
	Si	CIS	sc-Si	mc-Si	a-Si
Concrete	1193	1193	987.9	1192.9	1919.8
Steel	270.5	270.5	210.1	285.3	410.1
Copper	38.6	0.0435	13.6	16.3	26.1
Aluminum		0.054	10.1	0.1	0.1
Rubber/Plastic	7.8	7.8	10.1	11.1	9.4
SiO$_2$ (quartz pebble)	361.3				
Wood	186.9				
Glass	101.4				
Coal	74.7				
Petrol coke	49.8				
HCl (100%)	485.5				
SiC	11.3				
H$_2$	16.7				
N$_2$	930.1				
O$_2$	38.7				
Argon		16.2			
Molybdenum		0.2			
Indium		0.0785			
Selenium		0.325			

Table A10. Optical refractive indices for different wavelengths λ and $T = 20\text{–}25\ °C$

Material	λ in nm	n	Reference
Glass (Optiwhite®)	400	1.537	Flachglas 1989
	550	1.522	
	600	1.523	
	800	1.517	
	1000	1.514	
EVA (Elvax® 150) foil after lamination process A-9918 (20 min at 149 °C)	400	1.49	Gueris 1991
	600	1.47	
	800	1.45	
	1000	1.44	
	1200	1.43	
TiO_2 on substrate at 300 °C	400	2.73	Jellison 1985
	550	2.43	
	600	2.39	
	800	2.3	
	1000	2.27	
Silicon	354	5.61	CRC 1994
	400	5.57	Palik 1985
	496	4.32	CRC 1994
	600	3.95	Palik 1985
	729	3.75	CRC 1994
	800	3.69	Palik 1985
	1000	3.57	

Table A11. Absorption coefficients α at different wavelengths λ for $T=25\,°C$

Material	d in mm	λ in nm	α in m^{-1}	τ	Reference
Glass (Optiwhite®) chemically strengthened glass (CSG)	2	400	1.22	0.912	Flachglas 1989
		550	0.456	0.917	
		700	1.96	0.914	
		850	4.38	0.911	
		1000	5.31	0.910	
EVA (Elvax® 150) Foil after lamination A-9918 (20 min at 149°C)	0.5	390	323.20	0.787	Gueris 1991
		543	26.15	0.918	
		705	23.65	0.923	
		831	25.20	0.923	
		1026	25.75	0.925	

Table A12. Thicknesses of encapsulation materials applied

Thicknesses of the materials at the front side of a PV module in mm

Type of module	Glass	EVA	TiO$_2$	Si	Reference
Solarex IV	4.75	0.91	0.75 µm	0.305	Hoelscher 1981
SM 55	3.00	0.5	0.75 µm	0.450	Siemens 1991
PQ 40/50	2.00	0.5	0.75 µm	0.350	Telefunken 1991

Thicknesses of materials applied at the back side of a PV module in mm

Type of module	EVA	Fiberglass	Tedlar	Glass	Reference
Solarex IV	0.457	0.13 [1]	0.1	0	Hoelscher 1981
SM 55	0.5	0	0.150 [2]	0	Siemens 1991 Palmer 1992
PQ 40/50	0.5	0	0	2	Telefunken 1991

Comments:
[1] with second EVA layer, thickness also 0.457 mm
[2] Tedlar (37.5 µm) - Polyester (75 µm) - Tedlar (37.5 µm) - Laminate

Table A13. Thermal conductivity k of typical materials for PV systems

Material	T in °C	k in W (m K)$^{-1}$	References
Cell materials:			
Silicon	-23	191	CRC 2005
	27	148	CRC 2005
	77	119	CRC 2005
Germanium	-23	74.9	CRC 2005
	27	59.9	CRC 2005
	77	49.5	CRC 2005
GaAs		54	Heywang/Müller 1980
SiO$_2$	0	140	CRC 1994
	100	160	CRC 1994
ZnO	200	17	CRC 1994
MgO	100	36	CRC 1994
TiO$_2$	100	6.5	CRC 2005
TiO$_2$ (parallel to direction of crystal)	0	13	CRC 1994
	36	12.6	Crawford/Carson
	67	13.8	Crawford/Carson
TiO$_2$ (perpendicular to direction of crystal)	0	9	CRC 1994
	44	8.8	Crawford/Carson
	67	7.1	Crawford/Carson
Module/Encapsulation:			
Glass (ASG 13.80)		1.00	Garcia 1985
Flint glass	20	0.78	Kuchling 1995
Crown glass	20	1.07	Kuchling 1995
Window glass	20	0.81	Gieck 2005
Quartz glass	20	1.40	Baer *et al.* 1995
EVA		0.35	Garcia 1985
Elvax® 150	20	0.3461	Gueris (Du Pont) 1991
Elvax® 450	20	0.3375	Gueris (Du Pont) 1991

Material	T in °C	k in W (m K)$^{-1}$	References
Silicone (Semicosil® 912)	20	0.2	Wacker 1990
Polyurethane foam (PU)	20	0.06	CRC 1994
PE (Polyethylene)		0.5	Hornbogen 1994
	20	0.35	Baer et al. 1994
PMMA (Acrylic. Plexiglas®)	20	0.184	Baehr et al. 1995
PTFE (Teflon®)		0.24	Hornbogen 1994
Polyamide	20	0.29	Baer et al. 1994
Tedlar®		0.167	Garcia 1985
Support structure:			
Aluminum	-23	235	CRC 1994
	(20)	202	Khartchenko 2004
	27	237	CRC 1994
	77	240	CRC 1994
Cooper	-23	406	CRC 1994
	(20)	385	Khartchenko 2004
	27	401	CRC 1994
	77	396	CRC 1994
Iron	(20)	55	Khartchenko 2004
Steel	20	47	Mende/Simon 1981
	(20)	45	Khartchenko 2004
Stainless steel	0	14	CRC 2005
	100	16	CRC 2005
Cr-Ni-steel (X12CrNi18.8)	20	15	Baer et al. 1994
Cr-steel (X8Cr17)	20	25	Baer et al. 1994
Zinc	20	121	Baer et al. 1994
Tin	20	67	Baer et al. 1994
Concrete	20	1.45	Stöcker 1993
	(20)	0.8–1.73	Khartchenko 2004

Material	T in °C	k in W (m K)$^{-1}$	References
Ice	0	2.2	CRC 2005
Ceramics	(20)	1.7–2.9	Khartchenko 2004
Granite	20	2.90	Baer et al. 1994
Marble	20	2.80	Baer et al. 1994
Plaster	20	0.51	Baer et al. 1994
Aspahlt	20	0.06	CRC 2005
Wood: oak, beech	20	0.17	Stöcker 1993
Wood: spruce	20	0.14	
Water	0	0.5610	
	25	0.6071	CRC 1994
	50	0.6435	
Air (dry)	0	0.02454	
	20	0.0260	Gieck 2005
	100	0.0320	

Table A14. Emissivities of different materials

Material	Temp. in °C	ε	λ_e in nm	"Radiation coefficient" in $W/(m^2 K^4)$	Reference
Water	0	0.95			Holman 1990
	100	0.96			Holman 1990
Glass	20	0.94			Weast 1987
	22	0.94			Holman 1990
	34	0.81			Touloukian 1967
	90	0.88			Kuchling 1995
	100	0.91			Weast 1987
	110	0.82			Touloukian 1967
Aluminum (polished)	20	0.04		0.30	Kuchling 1995
Aluminum (raw)	20	0.06		0.41	Mende et al. 1981
Aluminum (oxidized)	20	0.30			Kuchling 1995
	200	0.11			Weast 1987
Stainless steel (polished)	100	0.07			Holman 1990
Stainless steel	25	0.60			Holman 1990
Steel (new)	100	0.08			Weast 1987
Steel (little oxide)	20	0.82			Weast 1987
Steel (lot of oxide)	50	0.88			Weast 1987
Steel (nickel plated)	20	0.11			Weast 1987
Nickel (non-oxidized)	25	0.05			Weast 1987
	100	0.06			Weast 1987
Zinc	28	0.23			Holman 1990
Cooper (oxidized)		0.76			Khartchenko 2004

Material	Temp. in °C	ε	λ_ε in nm	"Radiation coefficient" in $W/(m^2 K^4)$	Reference
Cooper (polished)		0.03			Khartchenko 2004
Silver	100	0.02			Weast 1987
Titan-dioxide	1,400	0.82	10,000		Sala 1986
	1,400	0.89	15,000		Sala 1986
Zinc-oxide		0.15	550		Weast 1987
Silicon	1,000	0.79	100		Sala 1986
	1,000	0.51	200		Sala 1986
Marble (light grey, polished)		0.95		5.37	Khartchenko 1995
Roofing felt	21	0.91			Holman 1990
Tile	20	0.93			Mende et al. 1981
		0.93			Khartchenko 2004
Concrete		0.88–0.97			Khartchenko 2004
Oak (planed)		0.91		5.16	Khartchenko 1995
Poly-carbonate	20	0.80			Kuchling 1982
black paint		0.95–0.97			Khartchenko 2004
org. plastics	20	0.90			Kuchling 1995
black body	all	1.0		5.67	Khartchenko 2004

Data for the "radiation coefficient" have been set-up to simplify calculations according to Khartchenko 1995, valid only for a small thermal range (50–100°C)

Table A15. Properties of Air (according to VDI 1994)

Temperature T in °C	ρ in kg m^{-3}	c_p in kJ (K kg)$^{-1}$	β in 10^{-3}/K	λ in 10^{-3} W m^{-2} K^{-1}	η in 10^6 Pa s	ν in 10^{-7} m^2/s	Pr
0	1.275	1.006	3.674	24.18	17.24	135.2	0.7179
10	1.230	1.007	3.543	24.94	17.74	144.2	0.7163
20	1.188	1.007	3.421	25.69	18.24	153.5	0.7148
30	1.149	1.007	3.307	26.43	18.72	163.0	0.7134
40	1.112	1.007	3.200	27.16	19.20	172.6	0.7122
60	1.045	1.009	3.007	28.60	20.14	192.7	0.7100
80	0.9859	1.010	2.836	30.01	21.05	213.5	0.7083
100	0.9329	1.012	2.683	31.39	21.94	235.1	0.7070

Table A16. Material composition of grid injection devices relevant for energy balance

Material	All components in kg/kW according to Johnson et al. 1997 (2.5 kW)	Inverter in kg/kW according to Frischknecht 1996 (500 kW)	Transformer in kg/kW according to Frischknecht 1996 (500 kW)	Control electronics in kg/kW according to Frischknecht 1996 (500 kW)
Steel	3.60	0.11	7.39	2.16
Aluminum	4.16	0.61	0.00	0.02
Copper	2.24	0.32	20.80	0.12
Silicon	-	0.00	0.00	0.00
Ceramics	0.08	0.21	0.00	0.00
LDPE	0.12	0.22	3.62	0.13
PP	-	0.04	0.00	0.00
PVC	0.04	0.06	0.00	0.00
Sulphide acid	-	0.00	0.00	0.02
Glass fiber	-	0.00	2.50	0.00
Not considered	-	0.09	0.35	0.14
Sum	10.24	1.62	36.40	2.60

Table A17. Official currency exchange rates used for conversion of data sets based on Euro (€) and German Marks (DM) to US-Dollar (US$) from 1975 to 2004 (yearly averages by the American Federal Bank and the Federal Reserve Bank of New York)

Year	Exchange rate DM per $US (yearly average)	Year	Exchange rate € per $US (yearly average)
1975	2.447	1995	(0.7383)
1976	2.538	1996	(0.7696)
1977	2.342	1997	(0.8870)
1978	2.031	1998	(0.8997)
1979	1.845	1999	0.9387
1980	1.798	2000	1.0835
1981	2.237	2001	1.1171
1982	2.415	2002	1.0578
1983	2.527	2003	0.8833
1984	2.819	2004	0.8040
1985	2.994	2005	0.8253
1986	2.215		
1987	1.827		
1988	1.746		
1989	1.881		
1990	1.636		
1991	1.653		
1992	1.561		
1993	1.643		
1994	1.633		
1995	1.444		
1996	1.541		
1997	1.735		
1998	1.760		

Fixed rate since 12/31/1998:
1 € (Euro) = 1.95583 DM (German Marks)
Values in brackets are based on the US$ vs. DM rate (left column)

Table A18. Fossil primary energy requirements and CO_2 emission for the generation of one MWh of electricity (data in parts by Mauch 1995)

Fuel (CO_2-intensity in kg/kg according to Table A2)	Coal (2.73)	Lignite (0.94)	Natural gas (2.29)	Diesel (3.27)	Sum CO_2
Aluminum exporting countries	55 kg	0	5 kg	2.2 kg	138.8 kg
Cooper exporting countries	187 kg	0	19 kg	5.5 kg	572.0 kg
West Germany	96 kg	228 kg	27 kg	2.2 kg	545.4 kg

Table A19a. Norms for PV systems and components (Source: IEC 8/2000)

IEC-Norm (year)	European Norm (year)	Title
IEC 60364-7-712		Electrical installations of buildings - Part 7-712: Requirements for special installations or locations - Photovoltaic power supply systems
IEC 60891 (1987-04)	DIN EN 60891 (1996-10)	Procedures for temperature and irradiance corrections to measured I-V characteristics of crystalline silicon photovoltaic devices
IEC 60891-am1 (1992-06)		Amendment 1 of IEC 60891
IEC 60904-1 (1987-12)	DIN EN 60904-1 (1995-04)	Photovoltaic devices, Part 1: Measurements of current-voltage characteristics
IEC 60904-2 (1989-05)	DIN EN 60904-2 (1995-04)	Photovoltaic devices, Part 2: Requirements for reference solar cells
IEC 60904-2-am1 (1998-02)		Amendment 1 of IEC 60904-2
IEC 60904-3 (1989-02)	DIN EN 60904-3 (1995-04)	Photovoltaic devices, Part 3: Measurement principles for terrestrial (PV) solar devices with reference spectral irradiance data
IEC 60904-5 (1993-10)	DIN EN 60904-5 (1996-07)	Photovoltaic devices, Part 5: Determination of the equivalent cell temperature (ECT) of photovoltaic (PV) devices by the open-circuit voltage method
IEC 60904-6 (1994-09)	DIN EN 60904-6 (1996-02)	Photovoltaic devices, Part 6: Requirements for reference solar modules
IEC 60904-6-am1 (1998-02)		Amendment 1 of IEC 60904-6
IEC 60904-7 (1998-03)		Photovoltaic devices, Part 7: Computation of spectral mismatch error introduced in the testing of a photovoltaic device

IEC 60904-8 (1998-02)		Photovoltaic devices, Part 8: Guidance for the measurement of spectral response of a photovoltaic (PV) device
IEC 60904-9 (1995-09)		Photovoltaic devices, Part 9: Solar simulator performance requirements
IEC 60904-10 (1998-10)		Photovoltaic devices, Part 10: Methods of linearity measurement:
IEC 61173 (1992-09)	DIN EN (IEC) 61173 (1996-10)	Overvoltage protection for photovoltaic (PV) power generating systems - Guide
IEC 61194 (1992-12)	EN (HD) 61194 (1996-07)	Characteristic parameters of stand-alone photovoltaic (PV) systems
IEC 61215 (1993-04)	DIN EN 61215 (1996-10)	Crystalline silicon terrestrial photovoltaic (PV) modules - Design qualification and type approval
IEC 61277 (1995-03)	E DIN IEC (CO) 19 (1989-10)	Terrestrial photovoltaic (PV) power generating systems - General and guide
IEC 61345 (1998-02)		UV test for photovoltaic (PV) modules
IEC 61427 (1999-11)		Secondary cells and batteries for solar photovoltaic energy systems - General requirements and methods of test
IEC 61646 (1996-11)		Thin-film terrestrial photovoltaic (PV) modules - Design qualification and type approval
IEC 61683 (1999-11)		Photovoltaic systems - Power conditioners - Procedure for measuring efficiency
IEC 61701 (1995-03)	E DIN IEC 1701 (1996-10)	Salt mist corrosion testing of photovolatic (PV) modules)
IEC 61702 (1995-03)		Rating of direct coupled photovoltaic (PV) pumping systems
IEC 61721 (1995-03)	E DIN IEC 1721 (1996-10)	Susceptibility of a photovoltaic (PV) module to accidental impact damage (resistance to impact test)
IEC 61724 (1998-11)		Photovoltaic system performance monitoring - Guidelines for measurement, data exchange and analysis
IEC 61725 (1997-05)		Analytical expression for daily solar profiles
IEC 61727 (1995-06)		Photovoltaic (PV) systems - Characteristics of the utility interface
IEC 61730-1		Photovoltaic module safety qualification - Part 1: Requirements for construction
IEC 61730-2		Photovoltaic module safety qualification - Part 2: Requirements for testing

IEC 61829 (1995-03)	Crystalline silicon photovoltaic (PV) array - On-site measurement of I-V characteristics
IEC/TR2 61836 (1997-10)	Solar photovoltaic energy systems - Terms and symbols
IEC 61836-2	Solar photovoltaic energy systems - Terms and symbols - Part 2
IEC 61853	Power and energy rating of photovoltaic (PV) modules
IEC 62078	Certification and accreditation program for photovoltaic (PV) components and systems - Guidelines for a total quality system
IEC 62108	Concentrator photovoltaic (PV) receivers and modules - Design qualification and type approval
IEC 62109	Electrical safety of static inverters and charge controllers for use in photovoltaic (PV) power systems
IEC 62116	Testing procedure - Islanding prevention measures for power conditioners used in grid connected photovoltaic (PV) power generation systems
IEC 62124	Photovoltaic (PV) stand-alone systems - Design qualification and type approval
IEC 62145	Crystalline silicon PV modules - Blank detail specification
proposals: DIN 40025	Data sheet and rating plate information for PV-modules

Table A19b. Norms for PV components and systems

IEC-Norm (year)	European Norm (year)	Title
1173 (1992-08)	DIN EN (IEC) 61173 (1996-10)	Surge protection for PV-power generating systems - Guide -
1194 (1992-12)	EN (HD) 61194 (1996-07)	Characteristics parameter of stand alone PV-systems
1277 (1995-02)	E DIN IEC (CO) 19 (1989-10)	Terrestrial PV-power generating systems - General and guide -
904-3 (1989-03)	DIN EN 60904-3 (1995-04)	Photovoltaic (PV) devices, Part 3: Measurement principles for terrestrial (PV) solar devices with reference spectral irradiance

904-5 (1993-11)	DIN EN 60904-5 (1996-07)	Photovoltaic (PV) devices, Part 5: Determination of the equivalent cell temperature (ECT) of PV-devices by the open circuit voltage method
904-6 (1994-09)	DIN EN 60904-6 (1996-02)	Photovoltaic (PV) devices, Part 6: Requirements for reference solar modules
904-7 (1995-09)		Photovoltaic (PV) devices, Part 7: Computation of spectral mismatch error introduced in the testing of a PV device
904-8 (1995-09)		Photovoltaic (PV) devices, Part 8: Guidance for the measurement of spectral response of a PV device
904-9 (1995-09)		Photovoltaic (PV) devices, Part 9: Solar simulator performance requirements
1215 (1993-4)	DIN EN 61215 (1996-10)	Crystalline silicon terrestrial PV-modules design qualification and design approval
1701 (1995-03)	E DIN IEC 1701 (1996-10)	Salt mist corrosion testing of photovolatic (PV) modules)
1721 (1995-03)	E DIN IEC 1721 (1996-10)	Susceptibility of a PV-module to accidental impact damage (resistance to impact test)
1886		Terminology used in the IEC PV-Norms

proposals:

1345		UV-test for PV-modules
1646		Thin-film terrestrial PV-modules design qualification and type approval
1730		Safety testing requirements for PV-modules
1798		Measurement of the linearity of a PV-device
1849		Design and type approval of PV-modules for marine environments
1853		Power and energy rating of PV modules
	DIN 40025	Data sheet and rating plate information for PV-modules

While the ISO 9000ff norms have been created for quality management purposes, the ISO 14000ff regulations have been set-up for environmental management later in May 1993 within a new technical committee named TC 207. This committee formed sub-committees (SCs) for the different subjects, which compiled the general requirements for the according standards. The detail work workgroup (WGs) have been nominated to specify formulations and values (such as in all ISO-projects). For the case of ISO TC 207 the following SCs have been set-up (see Table A20):

Table A20. The ISO 14000ff regulations

Name/description	Sub-Committee	ISO
Environmental Management System	SC 1	ISO 14000 - 14004
Environmental Auditing	SC 2	ISO 14010ff
Environmental Labelling	SC 3	ISO 14020ff
Environmental Performance Evaluation	SC 4	ISO 14031ff
Live Cycle Analysis	SC 5	ISO 14040ff

Beside these norms also other regulations as the *Swiss Eco-Audit* are used by the companies. This and other Eco-balances have the disadvantage that they are difficult to compare, while they are using different points of reference, e.g. the different borders for where the balances are carried out are leading to different results for the determination of the material- and energy flows. Efforts by a association called SPOLD ("Society for the Promotion of Life-Cycle Assessment Development"), attempt to use a common data format for all data acquired, see VDI-report 1328. To simplify comparison of different surveys, the publications are communicated via Internet since 1997.

Literature

Abid NC (1987) Contribution a l'étude de la production de froid a l'aide de capteurs a caloducs. Master thesis, Institute of Physics at the University of Constantine, Algeria

Allen MR, Scott PA., Mitchell JFB., Schnur R (2000) Quantif ying the uncertainty in forecasts of anthropogenic climate change. Nature 407: 617–620

Alsema EA, van Brummelen M (1995–1998) Minder CO_2 door PV. Vakgroep Natuurwentenschap en Samenleving, University of Utrecht, The Netherlands

Alsema EA, MJ de Wild-Scholten (2005) The real environmental impacts of crystalline silicon PV module: an analysis based on up-to-date manufacturers data. In: Proceedings of the 20th European Photovoltaic Solar Energy Conference and Exhibition, Barcelona

Archer CB (1980) Comments on "Calculating the Position of the Sun". Solar Energy 25: 91

Arrhenius S (1896) On the influence of carbonic acid in the air upon the temperature of the ground. Phil. Mag. Sci. 5: 237–276

Astronomical Almanac, 1997 Edition (1996) Particle Physics and Astronomy Research Council, London 1996; U.S. Govt. Printing Office, Washington D.C

Aulich H, Schulze F(1986) W. and B. Strake. Sonnenenergie, 6/86: 14 ff

Azzam RMA, Bashara NM (1987) Ellipsometry and Polarized Light. Amsterdam: Elsevier

Bach W (1996a) Energie und Klima. Spektrum der Wissenschaft: 30–36

Bach W (1996b) Weltbevölkerung, Energieverbrauch und Klimaschutz. Dossier: Klima und Energie: 24–33, Heidelberg: Spektrum der Wissenschaft Verlag.

BAM – Isecke B, Weltschev M, Heinz I (1990) Volkswirtschaftliche Verluste durch umweltverschmutzungsbedingte Materialschäden in der Bundesrepublik Deutschland. Bundesanstalt für Materialforschung und -prüfung (BAM), Institute for Environmental Protection, University of Dortmund

Becker W, Braun D (1990) Kunststoff-Handbuch. Vol 1; Die Kunststoffe. Hanser, Munich Vienna

Berz G (1996) Klimaänderung: mögliche Auswirkungen und Gegenmaßnahmen. Energiewirtschaftliche Tagesfragen. 46. Jg No. 7: 440–446

Bergmann, Schaefer; Niedrig, H. (Editor) (1993) Optik. 9th Edn. de Gruyter, Berlin New York

Beyer HG, Luther J, Steinberger-Willms R (1990) Zum Speicherbedarf in elektrischen Netzen bei hoher Einspeisung aus fluktuierenden erneuerbaren Energiequellen. Brennstoff, Wärme, Kraft 42: 430–435

Bloss WH, Pfisterer F (1992) Photovoltaische Systeme – Energiebilanz und CO_2-Reduktionspotential. VDI-Berichte 942: 71–87

BMWi – Bundesministerium für Wirtschaft (1996) Energiedaten'96 Nationale und Internationale Entwicklung. Editor: Referat für Öffentlichkeitsarbeit des Bundesministerium für Wirtschaft, 53107 Bonn (Germany)

Born M, Wolf E (1975) Principles of Optics (5th ed.). Pergamon, Oxford

Brauch HG (Ed) (1997) Energiepolitik: technische Entwicklung, politische Strategien, Handlungskonzepte zu erneuerbaren Energien und zur rationellen Energienutzung. Springer, Berlin Heidelberg New York

Bruton TM, Scott RDW, Nagle J P, Man MCM, Fackerall AD (1994) Re-Cycling of High Value, High Energy Content Components of Silicon PV Modules. In: Proceedings of the 12th European Photovoltaic Solar Energy Conference and Exhibition, Amsterdam. Vol.1, pp: 303–304

Campbell W (1997) Introduction to geomagnetic fields. Cambridge University Press, Cambridge (UK)

Cap F (1992) Graue Energie und der Treibhauseffekt. Österreichische Zeitschrift für Elektrizitätswirtschaft Vol 45 12: 507–519

Charlson RJ, Wigley TML (1996) Sulfat-Aerosole und Klimawandel. In: Dossier: Klima und Energie:. 74–81. Spektrum der Wissenschaft Verlag, Heidelberg

Charlson RJ, Schwartz SE, Hales JM. et al (1992) Climatic Forcing by Anthropogenic Aerosols In: Science, Vol 255: 423–430

Churchill SW (1977) A Comprehensive Correlating Equation for Laminar, Assisting, Forced and Free Convection. AIChE 10: 10–16

CRC – Lide DR. (ed.) (1994) Handbook of Chemistry and Physics. Boca Raton FL, (USA): CRC Press, Boca Raton, Florida

CRC – Lide DR (ed.) (2005) Handbook of Chemistry and Physics. Boca Raton FL, (USA): CRC Press, Boca Raton, Florida

Culp AW (1991) Principles of Energy Conversion. McGraw-Hill, New York

Day J, Johnson R, Gray D (1991) Markets, Sales and Distribution in the Photovoltaic Industry. In: Tutorial Notebook of the 22nd IEEE Photovoltaic Specialists Conference, Las Vegas, NV USA

DB (1995) Enquete-Kommission des 12. Deutschen Bundestages: Mehr Zukunft für die Erde: Nachhaltige Energiepolitik für dauerhaften Klimaschutz. Bonn: Economica Verlag

Dietz N (1991) Charakterisierung von Halbleitern für photovoltaische Anwendungen mit Hilfe der Brewsterwinkelspektroskopie. Ph.D. thesis at Hahn-Meitner-Institut Berlin, Berlin

DIN 5034 part 2 (1985) Tageslicht in Innenräumen; Normenausschuß Lichttechnik (FNL). Deutsches Institut für Normung e.V. Beuth, Berlin

Doka G, Frischknecht R, Hofstetter P, Knoepfel I, Suter P, Walder E, Dones, R (1995) Ökoinventare für Energiesysteme: Beispiel regenerative Energiesysteme. Brennstoff Wärme Kraft Vol. 47, 5:208–213

Dubbel; Eds.: Beitz W, Kütter KH (1995) Dubbel – Taschenbuch für den Maschinenbau. 18th edn., Springer, Berlin Heidelberg New York

Duffie JA, Beckmann WA (1974) Solar Energy Thermal Process. John Wiley & Sons, New York London Sydney Toronto

Ebersperger R (1995) Beispiele für Zurechnungsverfahren des Energieaufwands bei Entsorgung und Recycling von Produkten. VDI Berichte No. 1218: 11–31

Emery K, Burdick J, Caiyem Y, Dunlavy D, Field H, Kroposki B, Moriatry, T (1996) Temperature Dependence of Photovoltaic Cells, Modules and Systems. In: Proceedings of the 25th IEEE-PV-Specialists Conference., Washington D.C., USA, pp 1275–1278

Ewers HJ, Rennings K (1991) Die volkswirtschaftlichen Schäden eines Super-GAU's in Biblis. Zeitschrift für Umweltpolitik und Umweltrecht 4/91: 379–396

Faber M, Jöst F, Proops J, Wagenhals G (1996) Wirtschaftliche Aspekte des Kohlendioxid-Problems. In: Dossier: Klima und Energie: 43–51. Spektrum der Wissenschaft Verlag , Heidelberg

Faninger F (1991) Wege zur Reduktion energiebedingter Emissionen. In: ÖZE 44 (2): 43–63

Farkas I (1992) A Meteorological Model for Solar Engineering Applications. In: Proceedings of the Second World Renewable Energy Congress, Reading (UK).Vol. 5: 2731–2735

Fichtner (1986) Umweltvergleich von elektrischen mit anderen Heizungssystemen, Teil II. Ed.: Förderungsgesellschaft Technischer Ausbau e.v., Bonn

Fischedick M, Kaltschmitt M (1994) Stromversorgung unter Integration einer Elektrizitätserzeugung aus Windkraft, Photovoltaik und Biogas. In: Proceedings of the Ninth International Solar Forum, Stuttgart, Vol. 2: 1547–1554

Fischedick M, Barth V, Burdick B (1999) Stellungnahme zu dem Aufsatz "Kohlendioxid, Windenergienutzung und Klima" von Prof. Weigl. Report of Wuppertal Institute, Germany

Flachglas Solartechnik GmbH (1989) CSG-Opti-Solar, Chemically Strengthened Glass. Data sheet IOCR 2905. Cologne: Flachglas-AG

Flohn H (1989) Kann die Menschheit eine Klimaänderung verhindern? In: Energiewirtschaftliche Tagesfragen 39. Jg. 1–2: 42–46

Frisch von K (1965) Tanzsprache und Orientierung von Bienen. Springer, Berlin Heidelberg New York

Fischknecht R, Dones R, Hofstetter P, Knoepfel I, Zollinger E (1995) Ökoinventare für Energiesysteme. Grundlagen für den ökologischen Vergleich von Energiesystemen in Ökobilanzen für die Schweiz, im Auftrag des Bundesamtes für Energiewirtschaft und des Nationalen Energie-Forschungs-Fonds, ETH Zürich / Paul Scherrer Institute, 2nd Edn., Zürich (Switzerland)

Fischknecht R, Dones R, Hofstetter P, Knoepfel I, Zollinger E (1996) Ökoinventare für Energiesysteme. Laboratorium für Energiesysteme der ETH Zürich und Forschungsbereich 4 des Paul Scherrer Instituts, 3rd Edn., Zurich (Switzerland)

Fritsche U (1989) Emissionsmatrix für klimarelevante Schadstoffe in der BRD. Öko-Institut Darmstadt (Germany)

Fritsche U, Rausch L, Simon K-H (1993) Endbericht Gesamt-Emissions-Modell Integrierte Systeme (GEMIS), Hessisches Ministerium für Umwelt, Energie und Bundesangelegenheiten. Frankfurt a.M. (Germany)

Fuentes MK (1985) Thermal Characterization of Flat-Plate Photovoltaic Arrays. In: Proceedings of 18th IEEE Photovoltaic Specialists Conference, Las Vegas NV (USA), pp 203–207

Fujii T, Imura H (1972) Natural-Convection Heat Transfer from a Plate with Arbitrary Inclination. In: Int. Journal of Heat and Mass Transfer 15, pp. 755–757

Funck G (1985) Wärmeabführung bei Getrieben unter quasistationären Betriebsbedingungen Ph.D. thesis, Technical University of Munich (Germany)

Geiger B (1993) Energetische Lebenszyklusanalyse von Gebäuden. In: Kumulierte Energie- und Stoffbilanzen – ihre Bedeutung für Ökobilanzen. Meeting in Munich. VDI-Gesellschaft Energietechnik.VDI-Berichte Nr. 1093. VDI-Verlag, Düsseldorf

Gelen H (1994) Milieugerichte levencyclusanalyse van bulkmaterialien toegepast in zonnecelsystemen. Report 94068. Vakgroep NW&S, Universiteit Utrecht

GEMIS – Fritsche U, Leuchtner J, Matthes FC, Rausch L, Simon K-H (1992) Gesamt Emissions Modell Integrierter Systeme (GEMIS) Version 2.0 Endbericht. Öko-Institut & GH Kassel, Darmstadt Kassel

Gieck K (1989) Technische Formelsammlung. 19th Edn. Heilbronn: Gieck

Gieck K & R (2005) Technische Formelsammlung. 31st German Edn. (78th global Edn.). Gieck, Germering

Goetzberger A, Voß B, Knobloch J (1994) Sonnenenergie: Photovoltaik. B.G. Teubner, Stuttgart

Green M (1995) Silicon Solar Cells: Advanced Principles & Practice. Centre for Photovoltaic Systems and Devices, University of New South Wales, Sydney

Green M (2000) Photovoltaics: Technology overview. Energy Policy 28: 989–998

Grimm W-D, Schwarz U (1985) Naturwerksteine und ihre Verwitterung an Münchener Bauten und Denkmälern. Arbeitsheft 31. Bayrisches Landesamt für Denkmalpflege, Munich

Gueris C, Du Pont, European Technical Center (1991) Datasheet of ELVAX 150; Solar Transmittance and Reflectance, Refractive Index, Thermal Conductivity. Du Pont de Nemours, Le Grand-Saconnex (Switzerland)

Häberlin H, Röthilsberger HR (1993) Neue Photovoltaik-Wechselrichter im Test. Bulletin. Schweizerischer Elektrotechnischer Verein, Vol. 84, No. 10: 44–50

Hagedorn G (1989) Kumulierter Energieverbrauch und Erntefaktoren von Photovoltaik-Systemen. Energiewirtschaftliche Tagesfragen, Vol. 39 No. 11: 712–718

Hagedorn G (1990) CO_2-Reduktions-Potential photovoltaischer Systeme. Sonnenenergie 1/90 : 12–15

Hagedorn G, Hellriegel E (1992) Umweltrelevante Masseneinträge bei der Herstellung von Solarzellen. Forschungsstelle für Energiewirtschaft, Munich

Hanitsch R, Lorenz U, Petzold D (1986) Handbuchreihe Energieberatung/Energiemanagement. Eds.: Winje, D. and Hanitsch, R., Vol. V: Elektrische Energietechnik. Springer. Berlin Heidelberg, New York, Springer

Hantsche U, Hirtz W, Huber W, Kolb G (1991) Umweltvorsorgeprüfung bei Forschungsvorhaben am Beispiel Photovoltaik. Programmgruppe Systemforschung und technologische Entwicklung, Forschungsentrum Jülich GmbH (KFA), 3rd interim report, Jülich

Hantsche U (1993) Abschätzung des kumulierten Energieaufwandes und der damit verbundenen Emissionen zur Herstellung ausgewählter Baumaterialien. In: Kumulierte Energie- und Stoffbilanzen – ihre Bedeutung für Ökobilanzen. Meeting in Munich. VDI-Gesellschaft Energietechnik, VDI-Berichte Nr. 1093. VDI-Verlag Düsseldorf

Happoldt H, Oeding D (1978) Elektrische Kraftwerke und Netze. Fifth revised edn. Springer, Berlin Heidelberg New York

Hart GW, Raghuraman P(1982) Residential Photovoltaic-System Simulation: Thermal Aspects. Massachusetts Inst. of Technology, Lexington (USA)

Heywang W, Müller R (1980) Halbleiter-Elektronik. Winstel G. (Ed.), Vol. 10. Optoelektronik I. Springer, Berlin Heidelberg New York

Hoagland W(1996) Regenerative Energien: Sonnenenergie. In: Dossier: Klima und Energie, Spektrum der Wissenschaft. Heidelberg, pp 100ff

Hoelscher JF (1981) The Solarex IV Module. The Conf. Record of the 15th IEEE Photovoltaic Specialists Conference, Orlando FL (USA), pp 745–749

Höner R (1997) Wirtschaftlichkeitsanalyse der photovoltaischen Stromversorgung anhand von Energiebilanzen. Studienarbeit am Institut für Elektrische Energietechnik am Fachbereich Elektrotechnik der TU Berlin

Hoffmann VU, K Kiefer (1994) Das 1000-Dächer-Programm: Eine Zwischenbilanz. Sonnenenergie & Wärmetechnik 2/94: 22–97

Holman J P (1990) Heat Transfer. Mc Graw-Hill, New York
Homeyer O (1989) Soziale Kosten des Energieverbrauchs. Second revised and extended edn. Springer, Berlin Heidelberg New York
Homeyer O, Gärtner M (1992) Die Kosten der Klimaänderung – Bericht an die EU-Kommission. Fraunhofer-Gesellschaft, Karlsruhe (Germany)
Hornbogen E (1994) Werkstoffe. 6th edited and revised edn. Springer, Berlin Heidelberg New York
Huber W, Kolb G (1995) Life cycle Analysis of Silicon based Photovoltaic Systems. Solar Energy 54:153–163
Hummel F, Müh H, Wenisch R, Bystron K, Pfeifle R (1994) Wirkungsgrade und Netzrückwirkungen verschiedener Wechselrichter. In: Proceedings of the Ninth International Solar Forum, Stuttgart (Germany) Vol. 1, pp 566–573
Hütte – Akaedemischer Verein Hütte e.V. (Ed.) (2000) Grundlagen der Ingenieurwissenschaften. edited by H. Czichos, 31st revised and extended Edn. Springer, Berlin Heidelberg New York
Hütte – Akaedemischer Verein Hütte e. V. (Ed.) (2004) Die Grundlagen der Ingenieurwissenschaften. edited by H. Czichos and M. Hennecke, 32nd revised and extended Edn. Springer, Berlin Heidelberg New York
IPCC – Intergovernmental Panel on Climate Change (1990) The IPCC Scientific Assessment. Eds.: J. T. Houghton, G. J. Jenkins, J. J. Ephraums. Cambridge University Press, Cambridge, UK
IPCC – Intergovernmental Panel on Climate Change (1994) Radiative forcing of climate change. Report to IPCC from the Scientific Assessment Working Group, WMO/UNEP
IPCC – Intergovernmental Panel on Climate Change (2001) Third Assessment Report (TAR)
Jantsch M, Schmid H, J Schmid (1993) Einfluß von Qualität der Systemkomponenten auf die Energiebilanz von Photovoltaik-Anlagen. In: Proceedings of the Eighth Symposium on Photovoltaic Solar Energy, Staffelstein (Germany), pp 209–219
Jellison GE, Wood RF (1985) Antireflection Coatings for Planar Silicon Solar Cells. Solid State Division, Oak Ridge National Laboratory, Oak Ridge, TN (USA)
Jensen Ch (1961) Die Polarisation des Himmelslichts. In: Handbuch der Geophysik, Vol. VIII, Chapter 9, Eds.: Linke F, Möller F, Gebrüder Borntraeger, Berlin, pp 1942–1961
Johnson AJ, Outhred HR, Watt M (1997) An Energy Analysis of Inverters for Grid-Connected Photovoltaic Systems. In: Proceedings of the 14th European Photovoltaic Conference, Barcelona (Spain), pp 2194–2197
Joos F, Sarmiento JL (1995) Der Anstieg des atmosphärischen Kohlendioxids. Physikalische Blätter 51, Nr. 5: 405–411
Kambezidis HD, Papanikolaou NS (1990) Solar Position and Atmospheric Refraction. Solar Energy 44:143-144
Kaltschmitt M, Voß A (1991) Leistungseffekte einer Stromerzeugung aus Windkraft und Solarstrahlung. Elektrizitätswirtschaft 90 8:365–371
Kaltschmitt M (1994) Erneuerbare Energieträger im Kontext des Energiesystems der Bundesrepublik Deutschland. In: Proceedings of the Ninth International Solar Forum, Stuttgart (Germany), Vol.2, pp 1709–1716
Kaltschmitt M, Wiese A (1995a) Erneuerbare Energien: Systemtechnik, Wirtschaftlichkeit, Umweltaspekte. Springer, Berlin Heidelberg New York

Kaltschmitt M, Fischedick M (1995) Wind- und Solarstrom im Kraftwerksverbund – Möglichkeiten und Grenzen. C. F. Müller, Heidelberg

Kaltschmitt M, Wiese A (Ed.) (1997) Erneuerbare Energien: Systemtechnik, Wirtschaftlichkeit, Umweltaspekte. 2nd edn. Springer, Berlin Heidelberg New York

Kaltschmitt M, Wiese A, Streicher W (Ed.) (2003) Erneuerbare Energien: Systemtechnik, Wirtschaftlichkeit, Umweltaspekte. 3rd revised edn. Springer, Berlin Heidelberg New York

Khartchenko N (1995) Thermische Solaranlagen. Verlag für Wissenschaft und Forschung – VWF, 1st Edn. Berlin

Khartchenko N (2004) Thermische Solaranlagen. 2nd revised and extended edn. Verlag für Wissenschaft und Forschung – VWF, Berlin

Kaufman YJ. Dah-Ming Chou (1993) Model Simulations of the Competing Climate Effects of SO_2 and CO_2. J of Climate. 6, No. 7:1241–1252

Kayne J, Outhred H, Sørensen B (1992) System Aspects of Grid-Connected Photovoltaic Power Systems. In: Proceedings of the 11th European Photovoltaic Solar Energy Conference and Exhibition, Montreux (Switzerland), pp 1195–1198

Keeling CD, Whorf TP (2004) Atmospheric CO_2 concentrations (ppmv) derived from in situ air samples collected at Mauna Loa Observatory, Hawaii. Carbon Dioxide Research Group, Scripps Institution of Oceanography (SIO); University of California

Keoleian GA, Lewis GM (1997) Application of life-cycle analysis to photovoltaic module design. Progress in Photovoltaics 5:287-300

Kern W, Tracy E (1980) Titanium Dioxide Antireflection Coating for Silicon Solar Cells by Spray Deposition. RCA Review 41:133–180

Kiefer K, T Erge (1994) Ergebnisse der Auswertung des 1000-Dächer-Programms. In: Proceedings of the Ninth International Solar Forum Vol. 1, Stuttgart (Germany), pp 443–450

Kiehl JT, Briegleb BP (1993) The Relative Roles of Sulfate Aerosols and Greenhouse Gases in Climate Forcing. In: Science, 260: 311–314

King DL, Eckert PE (1996) Characterizing (Rating) the Performance of Large Photovoltaic Arrays for All Operating Conditions. In:Proceedings of the 25th IEEE-PV-Specialists Conference, Washington D.C. (USA), pp 1385–1388

Kleemann M, Meliß, M (1993) Regenerative Energiequellen. Second revised Edn. Springer, Berlin Heidelberg New York

Klein MV, Furtak TE (1988) Optik (Translation by A. Dorsel and T. Hellmuth). Springer, Berlin Heidelberg New York

Kleiss G (1997) Energetische Bewertung von photovoltaischen Modulen auf der Grundlage der Jahreswirkungsgradmethode. Köster, Berlin

Knaupp W (1996) Operation Behaviour of Roof Installed Photovoltaic Modules. In:Proceedings of the 25th IEEE-PV-Specialists Conference, Washington D.C., USA, pp 1445–1448

Krauter S, Hanitsch R (1990) The Influence of the Capsulation on the Efficiency of PV-Modules. In: Proceedings of the 1st World Renewable Energy Congress, Reading (UK), Vol. 1, pp 141–144

Krauter S, Hanitsch R, Strauß Ph (1991) Simulation-Program for Selecting Efficiency Improving Strategies of PV-Module-Encapsulations under Operating

Conditions. In: Proceedings of "Renewable Energy Sources '91" Int. Conf., Prague (CFSR), Vol. III, pp 48–53

Krauter S, Hanitsch R (1992) Improvement of PV-Performance by Partly Structured Surfaces. Proceedings of the 6th Photovoltaic Science and Engineering Conference, New Delhi (India), pp 1110

Krauter S, Hanitsch R, Diwisch N (1992d) Measuring the Heat-Transfer Coefficient of PV-Modules. In: Proceedings of the 2nd World Renewable Energy Congress, Reading (UK), pp 562–566

Krauter S, Hanitsch R (1992e) Performance of a Partly Structured Surface at a PV-Module. In: Proceedings of the 11th European Photovoltaic Solar Energy Conference an Exhibition, Montreux (Switzerland), pp 1351–1354

Krauter S, Hanitsch R (1993a) Calculating the Influence of Skylight-Polarization on the Transmission of Encapsulations of PV-Modules. In: Proceedings of the Cairo International Conference of Renewable Energy Sources, Cairo (Egypt)

Krauter S, Hanitsch R (1993b) Optical and Thermal Parameters of PV-Module Encapsulation Improving Output Power. In: Proceedings of the ISES Solar World Congress, Budapest (Hungary), pp 249–254

Krauter S (1993c) Betriebsmodell der optischen thermischen und elektrischen Parameter von PV-Modulen. Köster, Berlin

Krauter S, Hanitsch R, Campbell P, Wenham SR (1994a) Optical Modelling, Simulation and Improvement of PV Module Encapsulation. In: Proceedings of the 12th European Photovoltaic Solar Energy Conference and Exhibition, Amsterdam, Vol. 2, pp 1198–1201

Krauter S, Hanitsch R, Strauß Ph (1994b) Einfluß optischer und thermischer Parameter auf den Tageswirkungsgrad von photovoltaischen Modulen. Ninth International Solar Forum, Berlin (Germany), Vol. 1, pp 495–502

Krauter S, Hanitsch R, Wenham SR (1994c) Simulation of Thermal and Optical Performance of PV Modules. In: Proceedings of the 3rd World Renewable Energy Congress, Reading (UK)

Krauter S, Hanitsch R, Moreira L (1996a) New Optical and Thermal Enhanced PV Modules Performing 12 % Better under True Module Rating Conditions. In: Proceedings of the 25th IEEE-PV-Specialists Conference, Washington D.C., USA, pp 1323–1326

Krauter S, Hanitsch R (1996b) Actual Optical and Thermal Performance of PV Modules. Solar Energy Materials and Solar Cells 41/42: 557–574

Krauter S (2004) Increased electrical yield via water flow over the front of photovoltaic panels. Solar Energy Materials and Solar Cells 82: 131–137.

Kuchling H (1995) Taschenbuch der Physik. 16th edition. Fachbuchverlag, Leipzig

LBST – Lutwig-Bölkow-Systemtechnik GmbH (1995). Studie "Solarfabrik '96" im Auftrag von Greenpeace e.V., Munich (Germany)

Leidner JR, Linton W (1991) L'impact des charactéristiques d'onduleurs sur la production d'énergie par des installations photovoltaiques. Bulletin Schweizerischer Elektrotechn. Verein, 82, No. 10: 37–44

Lehmann H, Raetz T(1995). Zukunftsenergien: Strategien einer neuen Energiepolitik. Birkhäuser, Berlin Basel Boston

Lenzen M (1999) Greenhouse Gas Analysis of solar-thermal electricity generation. Solar Energy 65, No. 6: 353-368

Levi L (1980) Applied Optics. A Guide to Optical System Design. Vol. 2. Wiley & Sons, New York Chichester Brisbane Toronto

Lewin B (1993a) CO_2-Emission von Energiesystemen zur Stromerzeugung unter Berücksichtigung der Energiewandlungsketten. Ph.D. thesis at the Department of Mining and Geosciences at the Technical University of Berlin

Lewin B (1993b) CO_2-Emission von Kraftwerken unter Berücksichtigung der vor- und nachgelagerten Energieketten. In: Kumulierte Energie- und Stoffbilanzen – ihre Bedeutung für Ökobilanzen. VDI-Berichte Nr. 1093. VDI-Verlag, Düsseldorf

Lelièvre J-F, Kaminski A, Boyeaux J-P, Monna R, Lemiti M (2005) Optical properties of PECVD and UVCVD SiN_x:H antireflection coatings for silicon solar cells. In: Proceedings of the 21st IEEE Photovoltaic Specialists Conference, pp 1111–1114

Linder H, Knodel H, Bläßer U, Danzer A, Knull U (1948/1977) Linder – Biologie. 18th total revised Edn. JB Metzlersche Verlagsbuchhandlung and Carl Ernst Poeschel Verlag, Stuttgart

Lobbes A (1997) Simulation der atmosphärischen Streuung und Himmelspolarisation von Sonnenstrahlung zur Bestimmung von Reflexionsverlusten an Solargeneratoren. Thesis at Department of Electrical Engineering, Technical University of Berlin

Lund PD, Pylkkänen T (1987) Fast Parametrization Procedure of Solar Cells. Helsinki University of Technology, Department of Technical Physics, Espoo (Finland)

Macagnan MH, Lorenzo E (1992) On the optimal size of inverters for grid-conneted PV systems. In: Proceedings of the 11th European Photovoltaic Solar Energy Conference, Montreux (Switzerland), pp 1167–1170

Mason NB, Bruton TM, Russell R (1995) Properties ans Performance of Coloured Solar Cells for Building Facades. In: Proceedings of the 13th European Photovoltaic Solar Energy Conference, Nice (France), pp 2218–2219

Mauch W (1995) Ganzheitliche energetische Bilanzierung von Kraftwerken. VDI Berichte Nr. 1218 pp 135–147. VDI-Verlag, Düsseldorf

Meadows DH, Meadows DL, Randers J, Behrens III, WW (1972) The Limits to Growth. Universe Books, New York

Mende D, Simon G (1981) Physik – Gleichungen und Tabellen. VEB Fachbuchverlag, Leipzig

Menges G (1977) Forschungsprogramm Wiederverwertung von Kunststoffabfällen. Ed.: Verband Kunststofferzeugende Industrie e.V. (VKE), Frankfurt/Main

Merker GP (1987) Konvektive Wärmeübertragung. Springer, Berlin Heidelberg New York

Mertens R, Nijs J, Van Overstraeten R, Palz W (1992) Summary of Panel Discussion. Technical Goals and Financial Means for PV Development. In: Proceedings of the 11th European Photovoltaic Solar Energy Conference, Montreux (Switzerland), pp 1009–1013

Mertens K (1997) Die weltweite Marktentwicklung der Photovoltaik. In: Sonnenenergie & Wärmetechnik 1/97:26–30

Moser R, Blum W (1993) Energiestatistik der Photovoltaikanlagen in der schweiz Ende 1992. Bulletin Schweizerischer Elektrotechnischer Verein, Vol. 84, No. 10: 11–15

Müller K, Block L, Neuendorf H (1994) Regenerative Energien – eine neues Dienstleistungsangebot der Berliner Energieversorgung (Auswertung „1000-

Dächer-Programm" und Projektvorstellung).In: Proceedings of the Ninth International Solar Forum, Stuttgart (Germany), Vol. 1, pp 451–458

Munich Re Group (2005) Topics Geo – Annual review natural catastrophes 2004. Munich Re, Munich

Münchener Rückversicherungsgesellschaft (1996) Jahresrückblick Naturkatastrophen 1995. In: Topics 1996, Munich (Germany)

Mullner AN, Roecker Ch, Bovin J (1997) Grid-connected PV installation: Comparison between land base and flat roof PV installations. In: 14th European Photovoltaic Conf., Barcelona (Spain), pp 889–892

Newinger M (1985) Einfluß anthropogener Aerosolteilchen auf den Strahlungshaushalt der Atmosphäre. Hamburger Geophysikalische Einzelschriften, Reihe A: Wissenschaftliche Abhandlungen, Vol. 73, Hamburg (Germany)

Nordmann T (1993) Behauptungen und Stellungnahmen zum Thema Photovoltaik. Bulletin. Schweizerischer Elektrotechnischer Verein, Vol. 84, No. 10: 41–43

Palik E D (1985) Handbook of Optical Constants of Solids.Academic Press, London

Palmer J (1992) Couraulds Performance Films. Andus, Chandler, AZ (USA)

Palz W, Zibetta H (1991) In: Solar Energy, Vol.10: 221 ff

PEC – Product Ecology Consulants: (1996) LCA methology. WWW, The Netherlands

Perez R, Seals R, Michalsky J (1993) An all-weather model for sky luminance distribution. Solar Energy 50: 235-245

Photocap – Solar Cell Encapsulants –Technical Guide (1996), Springborn Materials Science Corp., Enfield, CT (USA)

Posansky M (1991) Neue Möglichkeiten für die solare Stromgewinnung mit gebäudeintegrierten Solargeneratoren. Bulletin Schweizerischer Elektrotechnischer Verein, Vol. 82, No. 10: 49–50

Preu R, Kleiss G, Reiche K, Bücher K (1995) PV-Module Reflection Losses: Measurement, Simulation and Influence on Energy Yield and Performance Ratio. In: Proceedings of the 13th European Photovoltaic Solar Energy Conference, Nice (France), pp 1485–1488

Räuber A (1995) Entwicklungstendenzen der Photovoltaik. Technologie – Märkte – Forschungsförderung. In: Proceedings of the Tenth Symposium for Photovoltaic Solar Energy, Staffelstein (Germany), pp 27–33

Rauschenbach HS (1980) Solar Cell Array Design Handbook. chapter 9: Environments and their effects. Van Nostrand Reinhold, New York

Real M, Spreng D (1991) Energieaufwand zur Herstellung von Solarzellen – Besprechung eines Forschungsberichtes. Bulletin. Schweizerischer Elektrotechnischer Verein, 82, No. 10: 11–15

Reiche K, Kleiss G, Bücher K (1994) Energetische Bewertung von Photovoltaik-Modulen unter realistischen Bezugsbedingungen. In: Proceedings of the Ninth International Solar Forum, Stuttgart (Germany), Vol.1, pp 471–478

Reichert T (1996) Ermittlung des kumulierten Energieaufwands bei der Herstellung einer neuartigen kristallinen Silizium-Dünnschicht-Solarzelle. Master thesis at University of Applied Sciences Munich, Department of Electrical Engineering in colaboration with Lutwig-Bölkow-Systemtechnik Ltd (LBST), Ottobrunn (Germany)

Richaud A (1994) Photovoltaic Commercial Modules: Which Product for Which Market. In: Proceedings of the 12th European Photovoltaic Solar Energy Conference and Exhibition, Vol. 1, pp 7–14

Sala A (1986) Radiant Properties of Materials. Tables of Radiant Values for Black Body and Real Materials. Elsevier, Amsterdam

Sandter W (1993)"Zum Bund-Länder-1000-Dächer-Photovoltaik-Programm", In: Stromdiskussion: Dokumente und Kommentare zur energiewirtschaftlichen und energiepolitischen Diskussion: Erneuerbare Energien. Ihre Nutzung durch die Elektrizitätswirtschaft. IZE, Frankfurt a. M.

Schaefer H (1993) Zur Definition des kumulierten Energieaufwandes (KEA) und seiner primärenergetischen Bewertung. In: Kumulierte Energie- und Stoffbilanzen – ihre Bedeutung für Ökobilanzen. Tagung in München, 30.11.–1.12.1993. VDI-Gesellschaft Energietechnik, VDI-Berichte Nr. 1093. VDI-Verlag, Düsseldorf

Scheer H (1993) Sonnen-Strategie. Piper, München Zürich

Scheidewind P, Delaunay JJ, Rommel M (1994) Verbesserte Tageslichtsimulation durch gemessene und modellierte Leuchtdichteverteilung des Himmels. In: Proceedings of the Ninth International Solar Forum, Stuttgart (Germany), Vol. 2, pp 997–1004

Schmela M (2000) Looking for a grower? – Market survey of crystal growth equipment. Photon International, February: 30–35

Schmid J (1988) Photovoltaik – Direktumwandlung von Sonnenlicht in Strom. Verlag TÜV Rheinland, Cologne

Schmid J (Ed) (1994) Photovoltaik: Strom aus der Sonne; Technologie, Wirtschaftlichkeit und Marktentwicklung. 3rd completely revised edition. Müller, Heidelberg

Schmid J (1988) Photovoltaik: Ein Leitfaden für die Praxis. Fachinformationszentrum Karlsruhe (Editor), 3rd completely revised edition. TÜV Rheinland, Cologne

Schmidt H, Sauer DU (1994) Praxisgerechte Modellierung und Abschätzung von Wechselrichter-Wirkungsgraden. Proceedings of the Ninth International Solar Forum, Stuttgart (Germany), Vol.1, pp 550–557

Schoedel S (1993) Photovoltaik: Grundlagen un Komponenten für Projektierung und Installation. 2nd edition. Pflaum, Munich

Scholze H (1988) Glas: Natur, Struktur und Eigenschaften. Springer, Berlin Heidelberg New York

Schönwiese C D (1995) Klimaänderungen: Daten, Analysen, Prognosen. Springer, Heidelberg Berlin New York

Schreitmüller KR, Kreuzburg J (1994) Zur ökonomischen und ökologischen Bewertung regnerativer Energieträger. In: Proceedings of the Ninth International Solar Forum, Stuttgart (Germany), Vol. 2, pp 1668–1675

Schulze Th, Weber C, Fahl U, Voss A (1992) Grundlagenuntersuchung zum Energiebedarf und seinen Bestimmungsfaktoren – 1. Zwischenbericht zum Forschungsbereich III: Rationelle Energieanwendung und Energiebedarfsanalysen. University of Stuttgart, Institute for Energy Economy and Rational Energy Use

Shabana MM, Namour T (1990) Optimum Thickness of Solar Module Front Layers for Maximum Power Output. In: Proceedings of the 1st Word Renewable Energy Congress Vol.1, Reading (UK), pp 141-144

Siemens Solar Ltd. (1991) Dicken der Einkapselungsmaterialien eines SM 55 PV-Modules. Fax, Munich (Germany)
Sissine F (1994) Renewable Energy: A New National Commitment? CRS Issue Letter IB93063, 10.2. CRS, Washington DC
Sjerps-Koomen EA, Alsema EA, Turkenburg WC (1996) A Simple Model for PV-Module Reflection Losses under Field Conditions. Solar Energy 57, No. 6: 421–432
Staiß F, Bönisch H, Mößlein J, Pfisterer F, Stellbogen D (1994) Die Bedeutung der Photovoltaik für eine Klimaverträgliche Energieversorgung in Baden-Württemberg. In: Proceedings of the Ninth International Solar Forum, Stuttgart (Germany), Vol.1, pp 465–470
Stelzer T, Wiese A (1994) Ganzheitliche Bilanzierung der Stromerzeugung aus erneuerbaren Energieträgern. Proceedings of the Ninth International Solar Forum, Stuttgart (Germany), Vol. 2, pp 1636–1643
Stewart LH (1995) Untersuchungen von Recyclingpotentialen durch Ausnutzen von Altstoffverträglichkeiten. Master thesis at the Technical University of Berlin, Institute for Technical Environment Protection, Department of Waste Economy
Stöcker H (1993) Taschenbuch der Physik. Verlag Harri Deutsch, Frankfurt a. M, Thun
Strauß Ph, Onneken K, Krauter S, Hanitsch R (1994) Simulation Tool for Prediction and Optimization of Output Power Considering Thermal and Optical Parameters of PV Module Encapsulation. In: Proceedings of the 12th European Photovoltaic Solar Energy Conference and Exhibition, Amsterdam, Vol. 2, pp 1194–1197
Strippel M, von Oheimb R (1994) Erfahrungen mit netzfernen Photovoltaiksystemen in der Landwirtschaft. Proceedings of the Ninth International Solar Forum, Stuttgart (Germany), Vol. 1, pp 626–633
Spreng D (1993) Net-Energy Analysis and the Energy Requirements of Energy Systems. Praeger, New York
Spreng D, Doka G, Knoepfel I (1995) Graue Energie; Energiebilanzen von Energiesystemen. Teubner, 1995 – Zürich: vdf, Hochschulverlag. an der ETH Zürich, Stuttgart
Swinbank WC (1963) Long-Wave Radiation from Clear Skies. Quarterly Journal of the Royal Meteorological Society 89: 339–348
Touloukian YS (1967) Thermophysical Properties of High Temperature Solid Materials. Volume 1: Elements. Macmillan, New York
Unites Nations (1992) Integrated Environmental and Economic Accounting, Handbook of National Accounting. Interim Version, New York
Vallvé X, Serrasolses J (1994) Stand-Alone PV-Electrification in La Garroxta (Catalonia, Spain): A 50 kW$_p$ Programme by the Users. In: Proceedings of the 12th European Photovoltaic Solar Energy Conference and Exhibition, Amsterdam, Vol. 1, pp 465–468
van Engelenburg BCW, Alsema EA, Schropp, REI (1995) Recycling of a-Si Solar Cells. In: Proceedings of the 13th European Photovoltaic Solar Energy Conference and Exhibition, Nice (France),Vol.1, pp 296–299
Vaucher S (1993) Aspects écologiques de la production des cellules solaires en silicium amporphe. Institute de Microtechnique, Université de Neuchâtel (Switzerland)
VDI-Kommission Reinhaltung der Luft (1989) Fortschritte bei der thermischen, katalytischen und sorptiven Abgasreinigung. VDI, Düsseldorf (Germany)

VDI – Verband Deutscher Ingenieure (Ed) (1991) VDI-Wärmeatlas. 6th edition, VDI, Düsseldorf (Germany)

Voermans R, Hoppe W (1994) Photovoltaikanlage „Neurather See" – Erfahrungen der ersten Betriebsjahre. Proceedings of the Ninth International Solar Forum, Stuttgart (Germany), Vol. 1, pp 363–370

Voß A (1993) Sonne – mehr Hoffnungs- als Energieträger? Manuscript

Wagner H-J (1992) Umweltaspekte photovoltaischer Systeme. Forschungverbund Sonnenenergie: "Themen 92/93"

Wagner H-J (1995a) Energie und Emission von Solaranlagen.VDI-Fortschrittsberichte, Reihe 6, Nr. 325. VDI-Verlag, Düsseldorf

Wagner H-J, Brandt Th (1995b) Ermittlung des Primärenergieaufwandes zur Herstellung ausgewählter Werkstoffe. Report, University of Essen, Essen (Germany)

Wagner H-J (1996) Energieketten von A bis Z - Erntefaktor und energetische Amortisationszeit. Elektrizitätswirtschaft, Jg. 95,Vol. 8, pp 448–456

Wagemann HG, Eschrich H (1994) Grundlagen der photovoltaischen Energiewandlung: Solarstrahlung, Halbleitereigenschaften und Solarzellenkonzepte. Teubner, Stuttgart

Walraven R (1978) Calculating the Position of the Sun. Solar Energy 20:393-397

Wambach K (1996) Untersuchungen zu den technischen Möglichkeiten der Verwertung und des Recyclings von Solarmodules auf Basis von kristallinem und amorphen Silizium. BMBF: Status report Photovoltaics, Bonn (Germany)

Weast RC (Ed.) (1987) Handbook of Chemistry & Physics. 68th Edn., 2nd reprint. CRC Press, Boca Raton FL (USA)

Welter Ph (1993) Die Energierücklaufzeit von Photovoltaikanlagen. Sonnenenergie & Wärmetechnik 4/93: 28–30

Wenham SR, Bowden S, Dickinson M, Largent R, Shaw N, Honsberg CB, Green MA, Smith P (1997) Low cost photovoltaic roof tile. Solar Energy Materials and Solar Cells Vol. 47, Issues 1-4: 325-337

Whiller A (1967) Design Factors Influencing Solar Collectors. Low Temperature Engineering for Solar Energy. ASHRAE, New York

Wiese A (1994a) Ausgleichseffekte, gesicherte Leistung und Speicherbedarf einer großtechnischen regenerativen Stromerzeugung in Deutschland. In: Proceedings of the Ninth International Solar Forum, Stuttgart (Germany), pp 1539–1546

Wiese A, Kaltschmitt M (1994b) CO_2-Substitutionspotential und Minderungskosten regenerativer Energieträger zur Stromerzeugung in Deutschland. In: Proceedings of the Ninth International Solar Forum, Stuttgart (Germany), Vol. 2, pp 1644–1651

Wiese A (1994c) Simulation und Analyse einer Stromerzeugung aus erneuerbaren Energieträgern in Deutschland. Ph.D. thesis at the Institute for Energy Economy and Rational Use of Energy, University of Stuttgart (Germany).

Wilk H (1994) Erste Ergebnisse aus dem österreichischen 200 kW Photovoltaik Breitentest. Proceedings of the Ninth International Solar Forum, Stuttgart (Germany), Vol.1, pp 371–381

Wilkinson BJ (1983) The Effect of Atmospheric Refraction on the Solar Azimuth. Solar Energy 30:295

Worrell E et al (1994) New gross energy requirement figures for materials production. Energy 19(6): 627–640

Zimmermann W (1995) Einstrahlungsabhängigkeit des Parallelwiderstandes von Solarzellen. In: Proceedings of the Tenth Symposium Photovoltaic Solar Energy, Staffelstein (Germany), pp 643–647

Zittel W, Baumann A (1992) Ökologische Belastungen durch solare Stromversorgung im Vergleich zu konventionellen Systemen. Proceedings of the Eighth International Solar Forum, Berlin (Germany), Vol. 2, pp 646–652

Printing: Krips bv, Meppel
Binding: Stürtz, Würzburg